AK47 Assault Rifle

AK47 Assault Rifle

THE REAL WEAPON OF MASS DESTRUCTION

NIGEL BENNETT

SPELLMOUNT

First published 2010

Spellmount Publishers, a division of The History Press
The Mill, Brimscombe Port
Stroud, Gloucestershire, GL5 2QG
www.thehistorypress.co.uk

© Nigel Bennett, 2010

The right of Nigel Bennett to be identified as the Author
of this work has been asserted in accordance with the
Copyrights, Designs and Patents Act 1988.

All rights reserved. No part of this book may be reprinted
or reproduced or utilised in any form or by any electronic,
mechanical or other means, now known or hereafter invented,
including photocopying and recording, or in any information
storage or retrieval system, without the permission in writing
from the Publishers.

British Library Cataloguing in Publication Data.
A catalogue record for this book is available from the British Library.

ISBN 978 0 7524 5389 7

Typesetting and origination by The History Press
Printed in Great Britain

CONTENTS

	Acknowledgements	6
	Introduction	7
Chapter One	The Creation of a Legend	13
Chapter Two	The Basics: AK47, AKM and AK74	23
Chapter Three	Inside the AK47	47
Chapter Four	How to Use It	63
Chapter Five	Worldwide	73
Chapter Six	Sniper Variations	133
Chapter Seven	Ammunition and Magazines	141
	Summary	147
	Further Reading	149
	Index	151

ACKNOWLEDGEMENTS

Richard Jones and all of the excellent staff at the National Firearms Centre Leeds UK for their outstanding help. All photographs are accredited to them unless otherwise stated. Steve Cook for pushing me to do this. Taff for believing in me. Nicola Bennett for her help in the preparation of the book proposal. Sean Holmes for his camera skills and Tony for his on lights. Thanks to my publisher – Spellmount of The History Press for letting me do this!

INTRODUCTION

The AK47 assault rifle is the world's most prolific and iconic weapon with over 80 million produced to date and production ongoing in at least eighteen countries around the globe as identified by the author. Turn on any news channel anywhere in the world and you see or hear the AK47 firing away in some desperate foreign conflict. Indeed, today the forces of the US, the UK and NATO face, not weapons of mass destruction (WMDs) in the form of nuclear, biological and chemical devices but the worlds 'real' weapon of mass destruction of the twentieth and twenty-first centuries – the AK47 assault rifle.

Invented and first put into production in the late 1940s by its brilliant designer Mikhail Kalashnikov, the AK47 and its successors the AKM and AK74 are simply the best assault rifle design ever put into the field. It has already armed determined men and women in defeating two superpowers. First the US in the bloody fight for Vietnam and south-east Asia, and then the Soviet Union, birthplace of the weapon, in Afghanistan. It continues to this day to be the weapon of choice for all fighters with a cause.

The AK47 assault rifle has been produced in greater numbers, featured in more films and books, served in more conflicts and reached a higher global recognition status than any other weapon in history. Why is this? How can a simple construction of steel and wood reach such heights?

For an assault rifle to be a success in a 'real world' environment it must be four things, simple, robust, reliable and accurate. When analysing the AK47, does it meet these four factors and how do other designs compare?

Let's look at what the criteria mean. First 'simple': using the weapon must be intuitive using easy to remember drills which under stressful conditions can be turned into quick effective actions. Simply put, can you train a soldier or indeed a civilian with no firearms experience to use the weapon in quick time? In addition, after basic drills have been acquired can they be retained and used when the operator is cold, wet, hungry, scared and crouched in the bottom of a water-logged trench. On the battlefield a soldier's weapon needs to be simple; anything else is only good for the firing range and the weapons trade shows.

Next of our four is 'robust' – it can't under any circumstance fall to bits to the point it wont work! Assault rifles get bashed against walls dropped off the backs of trucks and any number of abuses none of which can be allowed within the design to affect it from functioning as a weapon. It's got to be tough – very tough, in fact totally soldier proof and civilian proof. Well trained disciplined soldiers will whenever possible maintain their rifle but with the best will in the world the real conditions of the battle field puts pressures on the robust nature of any weapon to the extreme. To make matters worst most users of assault rifles are not the highly skilled troops but half trained locals 'up for a fight' or scared teenagers who for whatever social or economic reasons find themselves in some Army.

Thirdly it must be 'reliable'; every time you load it and pull the trigger it has to fire! It sounds obvious but assault rifles get stoppages, they jam, they don't always fire and that can cost a soldier his life.

Lastly is 'accurate'; accuracy means different things to different people but basically the assault rifle needs to be accurate within the tactical situation you are operating in. If the battle is taking place within a modern urban environment the ranges in which the enemy are being engaged are likely to be close, anywhere from 10 to 150 metres, and therefore the weapon needs to be accurate within this tactical environment.

Commentators have often used the phase 'in today's battlefield' when describing modern battle grounds and ranges of engagement. This implies the world didn't have cities or deserts to fight over in years gone by, but the fact is nothing has changed since the Second World War. Man continues to fight over the same ground be it cities, jungles or deserts and the development of the assault rifle came about in order to meet these multiple battlefield situations and ranges of engagement. To clarify this – the assault rifle needs to be capable of hitting a man-sized target at up to 300 metres maximum. Longer range targets will be engaged by machine guns, snipers and other support weapons; the simple truth is the killing with assault rifles is done from 10 – 300 metres and mainly under 100 metres and that's it.

So taking the four tests into consideration how does the legend that is the AK47 stand up? On the first three there is no room for debate. It's so simple the most ill educated and simple people of the world can use it with minimal training and certainly without formal small arms training as given in developed armies. It's robust – most AK47s and their successors the AKM and AK74 will soldier on for years and years. The author fired a Soviet-built AKM on a range in Vietnam in 2007 that was dated 1968 and it worked as if new! Even if some of the 'furniture' such as the butt stock falls off, the weapon will function as normal.

Reliability is excellent and the internal design of the weapon ensures it will fire and keep firing so long as it is fed with working ammunition; it takes a hell of a lot of mud, sand and heat to stop it. The AK47 is totally capable of fulfilling this criterion in any conditions and has proved it time and time again.

Accuracy is the only area for some debate, but depending on the model, place and time of manufacture and calibre of the weapon it's good enough for the battlefield and that is what matters. In addition it is worth commenting here on the calibre of the original weapon – 7.62x39mm. This round, the M43, is powerful and a fine example of the balance between a 'full sized' rifle round such as the NATO 7.62x51mm and smaller calibres. The M43 does the complete job; it has range and sheer 'knock down' effect on target, transmitting its relatively low velocity to massive 'thump'. Most of today's rifle designs are made in the standard NATO 5.56x45mm round, and there have been many complaints from battlefields around the world that it is not up to the job. No such complaints can be levied at the M43. This round of ammunition is as important to the success of the AK47 design as the weapon itself and adds to the list of elements that make the AK outstanding.

There is one other factor that comes into play and puts the AK47 in a league of its own, and that's sheer numbers. It's cheap to make and factories and machine tools for the production of the weapon can be obtained easily. China sells thousands upon thousands of cheap AK47 copies to anyone with hard currency or something of value to trade. Nearly every defence tradeshow around the world will feature a manufacturer selling a version the weapon – it's easy to buy new, second hand or tenth hand. It's the simple old law of supply and demand; millions of people want a gun, be it for a legitimate army of a recognized state or a mob hell bent on murder. If you want one or need one, someone will sell it to you.

Western small arms designers have spent millions of dollars trying to produce an assault rifle that meets these four tests and it has largely become a design world gone mad. The industry continues to re-invent the wheel time and time again producing new designs that give only the smallest of ergonomic improvements. The truth is that only the AK47 design meets all of these basic points and the challenge is simple – build the best quality AK47 you can and you've got the best assault rifle in the world. The Bulgarians have done just that and aimed at quality production of the AK47 design rather then re-inventing something and producing yet another assault rifle from scratch.

By concentrating on quality and keeping the basic design the same, Bulgaria has become the new home of the AK47, and while production and design continues in Russia, Bulgaria leads the way in the promotion of the 'best AK47 in the world'. China on the other hand leads the way in scale of production and a global sales drive, while emerging powers like Iran are beginning to take their place at the AK47 table.

Why make anything else? Iran is a good example of a country which has tried to produce and sell indigenous designs; in the end they built the AK. It sells, people want them and at the time of writing the most highly rated AK47 model in Afghanistan by the forces opposed to NATO is Iranian made. In the years of the Cold War the Soviet Union and the other communist countries of the world exported the weapon for political influence and as a show of support for like-minded counties and organisations. In today's world it's all about the money, be it hard cash or economic concessions. The world has changed and the AK47 has changed with it, the ultimate example of a true global product with no barriers.

The fact is until rifle ammunition has a revolutionary change away from the current 'cartridge', assault rifles and indeed all small arms design has reached it limits and probably did so by 1960. Mikhal Kalashnikov was once quoted as saying he would shake the hand of anyone who produces a better design then him – I don't believe Mikhail will be doing that for a very long time.

Since 1945 the threat of international nuclear war between the Soviet Union and the USA led to numerous wars by proxy fought out in the third world and developing nations. The fall of the Berlin Wall and the end of this phase in history has led on to an even more unstable world. From this change grew the era of the WMD myth which was born in order to justify the US's global vision of new threats and opportunities for influence. Countries opposed to the US in the last decades of the twentieth century and the beginning of the twenty-first were all supposed to have produced massive stockpiles of WMD or be developing nuclear weapons. But the facts do not support the myth and not one single US or Allied soldier has been

INTRODUCTION

killed by such weapons since the phrase was coined. The fact is that the total number killed by such weapons since 1945 can be measured in a few thousand, mainly in chemical attacks on civilian targets and some use in the Iraq–Iran war. The real weapon of mass destruction is the AK47 and it has killed in its millions.

A weapon that costs only a few hundred dollars has changed regimes, defeated professional armies, destroyed economies, brought massive political influence and freed the oppressed. It is all things to all men, bringing the power of good and evil down to a simple construction of steel and wood. The AK47 has been a tool of politics and an economic commodity transcending all environments within modern history; Mao's famous quote that 'all power comes from a barrel of a gun' might have been written for the AK47. Kalashnikov's design with its ease of use, simplicity of construction combined with cheap cost is one of the greatest inventions of man whether we like it or not – it's brilliant.

Although most people around the world have heard of the AK47 and a massive 'fun club' has grown up in North America and Europe of enthusiasts and collectors, it is all too often misunderstood and incorrectly reported. Much of what is written about it in book format and the general media is out of date, technically wrong and incomplete.

The author's fascination with the AK47 and its forebears started at an early age when he was given a book on the Vietnam War. The book contained an illustration of the Chinese-built weapon the Type 56-1, basically an AK47 with an under folding butt. Next came another book, this time on the Soviet Army which contained more pictures and drawings of the weapon and by this point the author was hooked.

This publication aims to simplify the subject, first by showing the three main types of weapon which are commonly referred to as the AK47 in most media. These three types are the AK47, the AKM and the AK74 and then the mass of variations produced around the world. The scope of this coverage is limited to the models directly copied from the AK47, AKM and AK74; other designs, heavily influenced but not direct 'copies' such as the Israeli Galil are therefore omitted. From time to time the term 'AK' is used in order to cover the generic models under one heading. Any errors and any model missing or unreported is down entirely to the author. The subject is vast but the message is simple; it is the most influential weapon man has today and is by any measure truly great.

I present to you the AK47.

CHAPTER ONE

THE CREATION OF A LEGEND

The origins of the AK47 can be found in the changing shape of conflict during the Second World War. Armies on both sides started the conflict mainly armed with rifle designs dating from the beginning of the twentieth century and some even before that. They were bolt action rifles with great accuracy and range but had limited magazine capacity and rate of fire. The economics of production were also different from what was to come. Skilled workers produced carefully crafted rifles to equip well trained peacetime armies with only the experience of the First World War as a reference point for infantry warfare. The sub-machine gun went some way to give the infantryman a degree of enhanced fire power and at very short range this was achieved, but at medium range and beyond they were near useless. The saving grace for many armies was the introduction of the 'light' machine gun; however the bolt action rifle was the infantry mainstay.

The Second World War brought manoeuvre to the battlefield combined with close range killing in cities, trenches and hedgerows. Firepower mixed with the requirement for cheap mass production became the goal. Mass mobilisation and conscription of the civilian population meant that training times and requirements changed. No longer could professional armies produce well trained soldiers skilled at marksmanship and 'rapid fire' from a bolt action weapon.

What was needed was 'firepower for the masses' produced quickly and cheaply by semi and unskilled workers. Nazi Germany was the first to embrace this thinking and fielded two excellent examples in the form of the

MG 42 machine gun and the STG44 assault rifle. Both were made using stamped metal methods with the emphasis on fire power for the troops. The STG44 when fielded was the first of a breed, bringing rate of fire, magazine capacity and a round of ammunition designed for the real battlefield conditions encountered.

Russia's forced entry into the war in 1941 rapidly showed the total inadequacies of the small arms in use by the Red Army and not until 1945 with the fielding of the SKS carbine – too late for the war – did the Red Army soldier get a credible weapon. The classic image of the 'Russian' soldier on the Eastern front would likely show him armed with the PPSH 41 sub machine gun. This automatic weapon was in itself a fine example of a gun of that type, reliable and robust but firing a pistol-type round of ammunition good only for very short range. The fact is that for most of the war Russian troops went into battle with a bolt action rifle or carbine with a magazine capacity of five rounds. Sheer guts, determination, weight of numbers and ruthless leadership had to do the rest. The Red Army infantry fought the war mainly with the following weapons for the individual soldier:

- Moisin Nagant M1891/30 Rifle in a 7.62 x 54mm calibre with a 5 round magazine
- Moisin Nagant M38 and M44 Carbine, 7.62 x 54mm calibre, 5 round magazine
- PPS 43 sub-machine gun, 7.62 x 25mm calibre, 35 round magazines
- PPSH 41 sub-machine gun, 7.62 x 25mm, 71 drum or 35 round magazine

Many other types were of course developed and fielded both before and during the war, including semi-automatic rifles none of which were practically successful. Only the PPSH 41 and PPS 43 sub-machine gun types met the needs of mass production with semi and unskilled labour.

The breakthrough in development came not in the design of a new rifle, sub-machine gun or carbine but in the design and development of a new round of ammunition, the 7.62 x 39mm or M43. In this the Soviets were ahead of the game; they had produced the so-called intermediate round. Drawing from the hard gained experience of war they concluded they needed a round that provided the hitting power of a full sized rifle round which operated within the battle ranges found on the 'real' battlefield. No longer would the skilled infantryman take single accurate shots at the enemy at 'long' range; instead the ill-trained conscript would fight close and medium range battles in ruined cities.

The intermediate round embraces the idea of a round of ammunition the same size in diameter as a full sized rifle round – 7.62mm – combined with a shorter overall length – 39mm. This gives on-target killing power, less recoil and less storage space needed within the weapon's overall design.

Even today this round of ammunition is the best example of its type and it totally does the job. While western small arms designers stick with 5.56 x 45mm as the main calibre of choice, this is mainly an outcome of economic production power rather than what is the best round of ammunition. Born out of the AR15 (M16) development and the weapon and calibres adoption by the US armed forces in the mid-1960s it has been deemed the 'standard' ever since. However in recent years some moves have been taken to look at larger calibres for infantry rifles such as 6.5mm and 6.8mm because the 'standard' 5.56mm round is just not good enough; simply put it does not have the killing power of a bigger round.

Manufactures of 5.56mm ammunition like to quote the benefits of low recoil in regard to accuracy and high muzzle velocity, which is true, but what puts the target down is the energy transmitted to the target– the sheer 'thump' into the human body – and a bigger round with slightly less velocity does this. The M43 7.62 x 39mm is still after all these years the best at killing and that's what a bullet is there to do. Firing a modern rifle in 5.56mm calibre on a firing range does give benefits in accuracy and low recoil, for the firer helps in this. Firing the larger M43 round is little different in the semi-automatic mode but the 'feel' the shooter gets is one of considerably more energy and power – it just feels like it should, like it means business.

The development of this new ammunition type was the catalyst for what was to come and without it the AK47 would not have been made. It was the higher command in the Red Army that sought the development of a weapon capable of using the round. The design of the AK47 itself was the work of Mikhail Kalashnikov and his team; the need for such a weapon was the work of the Soviet military.

The creation of the AK47 was influenced by two factors – one being a degree of chance and the other planning born out of need and experience. Many myths have grown up around the AK47's design, mainly centred on its designer Mikhail Kalashnikov. The most common is that Kalashnikov thought up the weapon while lying in a hospital bed recuperating from wounds received at the Battle of Bryansk. In fact he did not invent the AK47 during this period, but his chance wounding did give him the time to pursue his interest in small arms design (Kalashnikov had trained as a unit armourer before going into tanks). His first efforts were the design of a sub-machine gun and a carbine. Both designs were rejected by the authorities, but the experience gained in designing and presenting them was invaluable and Kalashnikov began to make a name for himself. Ironically his carbine design lost out to the excellent SKS carbine, but within a few years the SKS was swept aside as a poor historical relation of the AK47.

The second factor was the competition launched in 1945 by the Soviet military to find the best infantry weapon possible that could use the 7.62 x 39mm round. The competition was arranged so that designers and design teams had to submit their work anonymously. No 'famous' gun designer names were to be used so as not to influence the chairing selection committee. This method helped the little-known Kalashnikov and he set to work.

Although he is often thought of as a one-man team designing everything from scratch, in fact he was aided by a number of skilled engineers and draughtsmen. However, while they provided technical expertise, Kalashnikov was the driving force in bringing the design process together and laying out the required parameters for the new weapon. One man, wounded by chance in war, responding to a competition on a subject which had fascinated him for years was the right mix – the right man in the right place at the right time. Some commentators have claimed that Kalashnikov 'stole' the design from other foreign models such as the German STG44 – this is totally untrue. He was influenced by other designs such as the M1 Garand but the AK47 was all his and the framework for it was provided by the specifications and trials conducted by the Soviet authorities.

Kalashnikov's first design made it though the first cut of the competition and was selected for further evaluation along with a number of other competing designs in 1946. The trials associated with the competition continued to get harder and harder and Kalashnikov continued to improve his design as other competing designers and engineers fell by the wayside. One famous Soviet gun designer on seeing and inspecting Kalashnikov's gun was so impressed by it that he gave up! He declared to an audience that Kalashnikov's design was clearly the best, though was persuaded by others to continue only to be rejected at the next stage.

The tests that the designs had to pass included throwing loaded weapons into baths of dirty water and dragging them through sand. Weapons were dropped from a height onto a concrete floor and then fired to ensure they worked without malfunction. Kalashnikov's evolving design was the only one to succeed. The competition's managing committee didn't settle for anything less than practical weapon perfection. Kalashnikov changed many of the design aspects of his weapon during the trials and with each change he got closer to the end result. The 'powers that be' wanted a rifle that was simple, totally reliable in all conditions and in the new adopted 7.62 x 39mm calibre.

One wonders how many of the western rifle designs of the time would have faired in this competition and in particular how the early model AR15/M16 would have got one. Given that the M16 uses a direct gas system – a totally different design approach to the AK – putting it through the 'dirty water' bath trial would have been interesting to say the least. It is also interesting to note that the US Armed Forces' first purchase of the M16 was by the US Air Force, not by the Army or Marines, and was purchased without any serious competition or weapon trials against competing designs. It would appear that the Soviet Union in this respect was more democratic and open to competition then the United States of America!

The main trials were carried out between June and August in 1947 and about this time Kalashnikov moved from outside of Moscow to Izhevsk in the Urals, 1,000km east of the capital. In January 1948 Kalashnikov's design was selected. His rifle – the AK47 – was put into production. The Soviet Army conducted its own acceptance trials on the rifle but by 1949 it was adopted and taken into service and by the end of 1949 some 80,000 AK47 rifles had been produced.

Historic Timeline of the AK47
(Soviet Production Communist Era)

1945	Soviet authorities launch competition to find the 'best' assault rifle to meet specific criteria and to use the M43 7.62 x 39mm round
1948	Kalashnikov's design wins
1949	Kalashnikov's design is accepted by and enters Soviet Army service
1948–1949	AK47 type 1 in production and service
1949–1953	AK47 type 2 in production and service
1953–1959	AK47 type 3 in production and service
1956	First combat use in the suppression of the Hungarian Uprising
1959	AKM enters service and begins to replace the AK47 as the dominate model both in Soviet service and worldwide
1974	AK74 enters service, cambered for the new 5.45 x 39mm round

The introduction of the improved AKM model, the *Avtomat Kalashnikova Modernizirovanniya* (M for modernised), was the key landmark in the story of the AK. This model – lighter then the original AK47 and well made – that cemented the legend of the AK47. Most AKs in service today and most non-Soviet copies are of the AKM pattern or a hybrid between the AK47 and the AKM, as in the case of the Chinese Type 56 series. When the media of the world talk about the 'AK47' or the 'AK' it's mainly this model they are referring too. The interesting oddity to this however is the excellent Bulgarian production variants which owe their design heritage to the type 3 milled receiver AK47 rather then the lighter stamped metal AKMs.

The introduction of the AK74 with its unique 5.45 x 39mm M74 round was a Soviet response to what they saw as the new trend in smaller rifle ammunition. The US introduction of the 5.56 x 45mm round was of course the catalyst for its development. The aim was to produce a smaller round with higher velocity producing more accurate fire both in semi-automatic and full automatic firing modes. In addition the smaller lighter rounds would mean the average soldier could carry more ammunition for the same overall weight. In some respects the Soviet designers were successful; the AK74 does give a far greater muzzle velocity then AKs cambered for 7.62 x 39mm and the weapon is more accurate in all firing modes at comparable and greater ranges than an AKM. The design of the AK74 is greatly aided by the design of its muzzle compensator which is the best of its type in reducing muzzle climb – it's a design that really works.

This being said it is not the AK74 we see on the majority of battlefields today, but rather the 7.62 x 39mm AK47, AKM and the numerous worldwide copies of both. Why is this? Well it is down to what was called at the time 'the end of history' the end of the Soviet Union the end of the Soviet block and the end of Soviet political influence around the globe. The AK74 is a victim of political history. Entering Soviet Army service in 1974, the weapon wasn't produced or used by the Soviet Union's Warsaw Pact allies until the 1980s and in some cases such as Hungary not at all. Therefore when the Berlin Wall came down only Russia and the new countries created out of the former USSR were left holding the 5.45mm AK74 weapon. The rest still had 7.62mm AKs and some 5.45mm weapons. Hungary had none in service, the DDR (East Germany) had its taken out of service when it reunified with West Germany, Poland had a mixture of both, Czechoslovakia had none and Romania and Bulgaria had both ammunition types in service. Further afield outside of Europe no one had it (with the exception of those left behind in Afghanistan).

It is said that Kalashnikov himself disapproved of the introduction of the 5.45 x 39mm round and instead wanted to see further development of the 7.62mm round. It's a shame they didn't listen to him; North Korea (DPRK) is the only country today outside of the former USSR to have taken this round into service. The Soviet military wanted a 'modern' smaller calibre round which was superior to the US round and the AK was redesigned to fit it. Kalashnikov and his team did another brilliant job in making this happen but the decision to go down this route was not his. The experience of the Vietnam War where the 5.56mm M16 came head to head with the AK47, AKM and numerous variants all in 7.62mm should have pointed the way forward – keep to a bigger bullet. Every other AK-using country sticks to the 7.62mm round (or the 'world standard' NATO 5.56 x 45mm round in a small number of cases) and today this calibre is manufactured all over the world.

It is interesting to note that the biggest sales successes in recent years for Russian-produced weapons are all in the 7.62 x 39mm calibre. Following the fall of the Berlin Wall and the political shift in Russia from a communist to a capitalist regime sparked the Kalashnikov manufacturers to offer the weapon in 5.56mm round (thinking this was going to be a sales winner in 'western' markets) and the original 7.62mm in the form of the AK103. This variant however has proven to be the most favoured option; Venezuela purchased 100,000 models and the rights to manufacture it and the ammunition in country. Venezuela was not confined by any legacy of weapon systems in 5.56mm, nor was it obliged to use this calibre as part of any notion of needing a cartridge round the same as its 'allies' – they had free choice and chose the bigger round, the original M43.

The AK103 is basically an AK74 with plastic furniture re-worked backwards to take the 7.62 x 39mm round. Therefore reproducing the AKM model with a great muzzle compensator was Kalashnikov's view of the way forward in the early seventies when the smaller 5.45 x 39mm M74 round was imposed upon him and his design team. Along the same lines of thought the Chinese also selected a larger round to the Western 5.56mm & Soviet 5.45mm – they went with 5.8mm calibre giving greater range and hitting power then either of its rivals. The author got the opportunity to test fire the Chinese QBZ 95 assault rifle with its unique 5.8mm calibre and can testify that the 'feel' of greater power is certainly there combined with low recoil; it's an excellent round. In addition the Chinese round is optimised for penetration and with most armies fielding body armour today the designers have reflected the modern battlefield. Development work continues in China on this unique round of ammunition adding a new dimension to the world of integrated small arms design and standardisation of ammunition across infantry weapon systems. As such the Chinese are the only designers to have produced an assault rifle, light machine gun, sniper rifle and general purpose machine gun all in the same calibre.

The Soviets came close to this 'integrated system' of small arms design and calibre with the AK47/AKM and its light machine gun variant the RPK, all in 7.62 x 39mm calibre and all sharing the same basic design. This integrated system reduces training time and logistics.

The M43 round developed in the Second World War is still the best; it does the job at all battlefield ranges. History killed the 5.45 x 39mm round and confirmed the bigger round as king. The demise of the USSR and its allies therefore gave new life to the AK; all production outside the USSR was mainly aimed at AKMs and the surplus market from Eastern Europe was massive combined with Chinese production increasing and new markets where once the USSR dominated. With the Soviet Union out of the way other producers could have a field day and in some cases the transformation from communist states to western 'free market'-orientated economies meant that the best export some countries had in the 1990s was the AK.

War made the weapon famous, bringing it to the attention of the world's media and military alike. Some have tried to demonise it as a weapon of terror but the truth is more complex. It was first taken into combat in 1956 during the Hungarian Uprising when elements of the 'intervening' Soviet forces were equipped with it. Some of these fell into use by the Hungarians fighting for freedom and a small number ended up in the hands of the US military for evaluation.

The heroic defiance of the Hungarian people in the unequal and bloody fight for their national independence ended in failure and the first outing of the AK47 went largely unnoticed. US military evaluators and other western 'experts' largely overlooked the new infantry weapon as no more than a big sub machine gun – inaccurate and limited in application.

Western small arms designers and planners were still stuck in the mindset of a full sized rifle firing a full sized round of ammunition of the type used in both world wars. US planners started with the 7.62 x 51mm round and imposed this design on its NATO allies as a 'standard' then subsequently did a complete u-turn and adopted the 5.56mm round in 1965. No one in the west took up the mid way point and all overlooked the design of the AK47. The Vietnam War changed all that.

War is hell and the AK was built for hell; the war in Vietnam was an infantry war fought at close range in jungles, paddy fields, hills, trenches and cities – in fact every imaginable terrain and all of it hot and very wet. There was no harder test on a weapon. The forces of the National Liberation Front (NLF) other wise known as the Viet Cong in the west plus the regular forces on the North Vietnamese Army (NVA) used many small arms types, but they are most identified with the AK.

Training time for the communist forces were limited and the conditions the soldiers and their equipment had to bear was tough. Combat between the opposing forces when it came was brutal and the automatic fire power of the AK with its 'full size' bullet did the job. Firing through bush or into trenches and buildings needed the 7.62 x 39mm bullet and it showed the 5.56mm M16 in its early version to be totally lacking.

The soldiers who had to use the M16 could clearly see the benefits of the AK but national pride and the military industrial complex behind it pushed on with the M16 and its smaller round. The soldiers who had the AKs on the other hand wanted for nothing else – they knew they had a battle winner built for war.

Nearly every conflict since the mid-1960s has featured the AK in some capacity or other. The Cold War stand off between the US and the Soviet Union fuelled this at first with the Soviets and her allies happy to supply the weapon to anyone who claimed to be on 'their' side of the political divide at little or no cost. If you were fighting a war in the 1960s or 1970s and claimed to be socialist you could get crate after crate of AKs. However in a twist of fate it was Afghanistan in the 1980s that proved to be the other great battleground for the AK in the same way that the conflict in Vietnam had been in the 1960s and early 1970s.

The irony of this was that the Soviets themselves that were on the receiving end of there own weapon. When the Soviet Army invaded Afghanistan in 1979 the resistance was armed with a massive and bizarre mixture of weapons. This mixture included bolt action weapons dating from the turn of the century but it wasn't long before the AK became the weapon most associated with the Afghan fighters.

The US – seeing the opportunity to hit the Soviets hard in the same way they had been hit and humiliated in Vietnam – flooded the country with AKs, many from China or bought second hand from all over the world. It became an AK 'free for all', no questions asked. However, the same weapons are now being used against their suppliers – the US and its allies – by Taliban insurgents.

By the 1980s the Soviet Army was mainly armed with the 5.45mm AK74 while its enemy had the older 7.62mm versions. At first western intelligence and indeed the media was intrigued by the new calibre of the AK74. Some even dubbed it the 'magic' bullet owing to its unique design with its hollow tip which bent on impact. However it soon became clear that in the grim mountain battles the extra hit of the bigger, original AK round still proved its worth.

Many writers have dwelt upon the past conflicts involving the AK – Vietnam, the wars of the Middle East, Afghanistan in the 1980s – but it is today's conflicts that matter now and they are very different. Old style east/west politics and related conflicts have been replaced by nationalism, religious divides and economic imbalances. The modern world's conflicts are more akin to the wars of the nineteenth and early twentieth century and it's a confused and complicated mess of carnage. At the time of writing more people die of violence on the Mexican border in 'drug wars' then in many current day conflicts involving 'armies' and the AKs they use are brought in illegally from the US – the marketplace for much of their drugs.

The fact is that the robust design of the AK and its simplicity of use makes it the staple diet of modern war. It can be used time and time again and re-sold to someone else. Only by cutting the weapon up into small pieces can you get rid of it and then someone else will make and sell more!

The AK has defeated two superpowers in one century, equipped freedom fighters and terrorists alike and fought in every major modern conflict conceived by man. History made the AK and the AK made history.

CHAPTER TWO

THE BASICS: AK47, AKM & AK74

AK47: The Legend

The AK47 assault rifle entered serial production in 1948 and joined the Soviet Army alongside the SKS carbine as the standard combat weapon of the Soviet soldier. For several years the SKS carbine remained in service, being phased out of front line Soviet Army use in the mid-1950s. The AK47 rifle is surrounded by myth and misinformation fuelled by the media. It has gone through three major design changes and contrary to popular belief it had a relatively short production life of just over ten years. Stories in newspapers and news reports would have us believe that once Kalashnikov invented the rifle it entered service and stayed in production as the same weapon design year after year – this is not so.

Three major design phases covered the AK47 in its short production life (1948–1959) referred to by most researchers as Type 1, Type 2 and Type 3. It was the Type 3 version which became the most influential, picked up and copied by others. The Type 1, only in production for a year, was a brilliant design let down by poor manufacturing techniques – basically it fell to pieces! The bodywork – known as the receiver – on the Type 1 was made from stamped steel welded into place. The techniques employed were poor and hard use in the field caused the rifle to break apart, hence the Soviets' quick change of production method to a milled receiver. This second version, the Type 2, was made by machining out a forged block of steel to make the bodywork of the receiver. This milling process produced a solid and robust receiver resistant to breakage.

The Soviets' change in production method is a testament to their abilities to invest and develop the design as soon as the problem was identified. AK47s are often referred to as being made from a single block of steel but this is only true of the later production models the Type 2 and Type 3, not the first production model.

The AKM – the most common version in service today is made from stamped steel just like the first version. It was the production techniques and the technology of stamping steels and milling steels that drove the design and as soon as the Soviets had improved the technology of stamped steel production they went back to it with the excellent AKM.

AK47 Type 1
- Stamped receiver – welded and riveted construction
- Smooth side to receiver – no 'dimple' or cut out above magazine well
- Stock, forearm and pistol grip made from unlaminated beech wood
- Rear sight graduated to 800 metres
- Flat sided construction of magazine – often referred to as 'slab sided'
- Production 1948–1949
- Weight – 4.3kg

AK47 Type 1 assault rifle 7.62 x 39mm – the original first production series rifle and the start of the legend.

AK47 Type 2
- Milled receiver machined from a forged steel block
- Rectangular 'cut outs' on either side of the receiver above the magazine well
- Bolt carrier guides machined directly into receiver
- Stock attached to the receiver via metal housing – often referred too as a 'boot'
- Production 1949–1953.
- Weight – 3.8kg

AK47 Type 3
- Milled receiver
- Stock attached directly to the receiver
- Selector lever strengthened
- Selector level improved with two thumb 'holds'
- Front sling position changed to outside of gas block
- Stock sling position moved to left side of stock
- Stock and forearm made from laminated wood
- Magazine produced with reinforced side rigid ribs
- Gas piston modified and made 'smooth' with no fluting
- Known as the '7.62mm Light Weight Kalashnikov Assault Rifle (AK)'
- Production 1953–1959

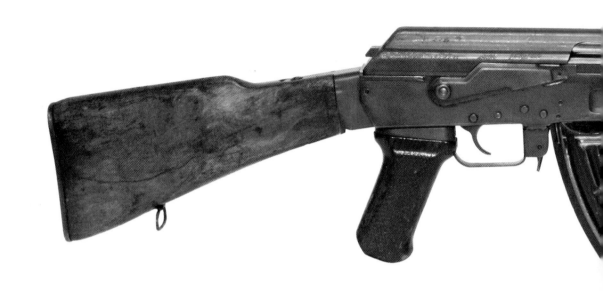

THE BASICS: AK47, AKM AND AK74

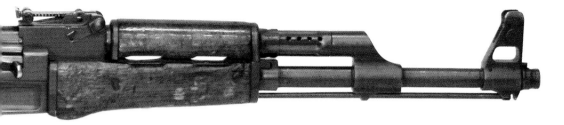

AK47 Type 2 7.62 x 39mm assault rifle. This second production type featured a milled receiver as opposed to the early stamped steel efforts of the Type 1. Note the 'shoe' connection between the rear of the receiver and the wooded stock. This feature is unique to the Type 2 and a good recognition point. Had the Type 1 remained in production and retained those early production methods it is unlikely that the AK would have gone on to become the most robust weapon in the world.

Soviet AKM

The final version of the AK47 and the version most associated with the rifle was only in Soviet production for six years! In one sense the AK47 was one big development trial gaining knowledge and skills in line with improvements in production methods leading to the ultimate design – the AKM. This version is sometimes referred to as the Type 4 AK47, but this designation is unhelpful – it's an AKM. What *is* of note here is the difference in speed of development between the AK47 leading to the AKM and later AK74 and the M16. The M16 remained basically unchanged until the improvements of the M16A2 were introduced in the 1980s – it took a free market capitalist country twice as long to sort out a rifle as the Soviet Union! Still today many military and civil commentators complain about many aspects of the M16 – no one complains about the AKM or indeed the Type 3 AK47. Its production outside of the Soviet Union and most notably in China combined with the design growth potential was its strength. The basic design was there it just needed to be lighter and easier to make.

Its heyday of use was of course the Vietnam War when it was fielded against the Americans. Soviet-built AK47s served alongside numerous Chinese Type 56 versions and provided the backbone of the North Vietnamese Army and regular National Liberation Front (Viet Cong) units.

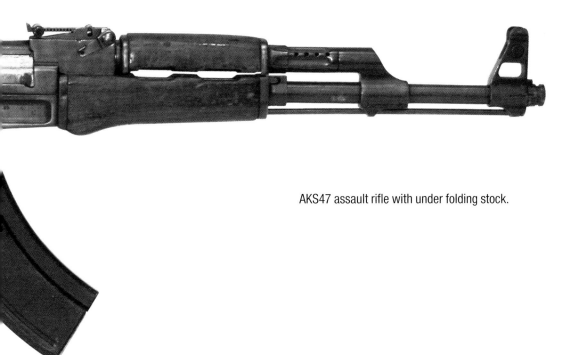

AKS47 assault rifle with under folding stock.

What the AK47 did was to provide the basis for the real 'AK' the AKM – light, tough, simple and well developed from the experience of the AK47. The Type 3 AK47 was a fine rifle and the milled receiver of this version still provides the inspiration for the excellent Bulgarian series of AK-type rifles. However, the AKM does it all and is the rifle we see today on the news, be it the Soviet original or one of the multitude of global copies. Many of the features of the Type 3 AK47 are still found within copies of the AKM around the world today. Hybrid features such as the arrangement of gas vent holes on the gas tube are evidence of this. Mixtures of AK47 features and AKM stamped steel receivers are common in non-Soviet production such as in the Chinese Type 56 and Hungarian AK-63 versions.

Because of their toughness and simplicity original AK47s are still widely encountered around the world today in war zones and will be for many years. Often parts are interchanged or remade to replace broken or worn out parts, such is the simplicity of the design. Use of steel and wood makes the process within the power of low-tech regimes and organisations. It's quite normal to find Type 3 AK47 rifles with re-made stocks or stocks taken from Chinese Type 56s and the whole package works just fine.

Much is written today about 'modular' small arms design with one weapon being able to meet many needs by changing barrels, calibres and

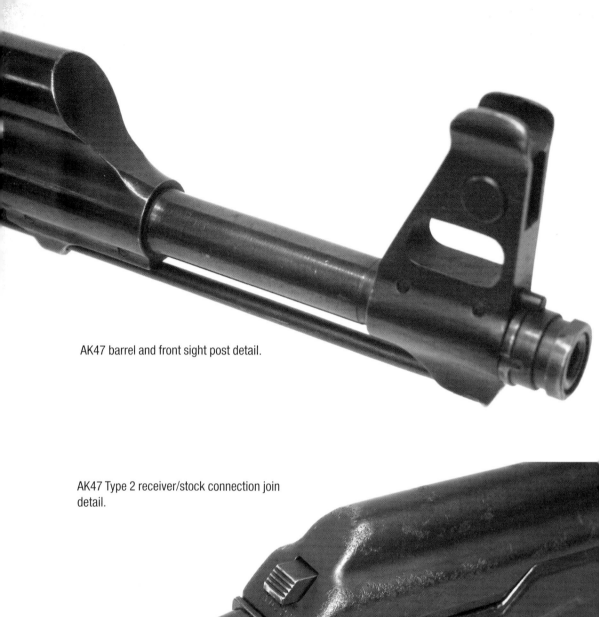

AK47 barrel and front sight post detail.

AK47 Type 2 receiver/stock connection join detail.

sight fittings. However, no-one has really come close to fulfilling this, with only the Chinese coming close with their range of 5.8mm weapons. It's the AK47 and its brilliant successor the AKM that does it – the combination of a simple internal design with few parts built in steel with a calibre that works on the battlefield does it – and all designed in the 1940s!

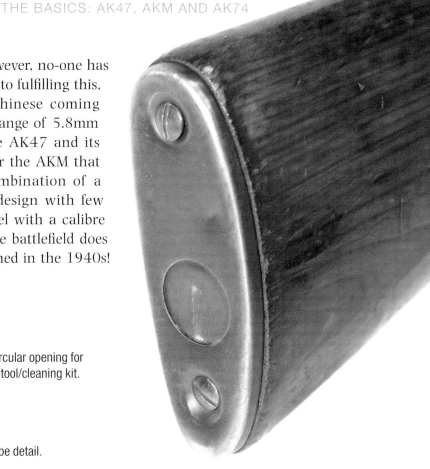

AK47 stock. Note the circular opening for storage of combination tool/cleaning kit.

AK47 barrel and gas tube detail.

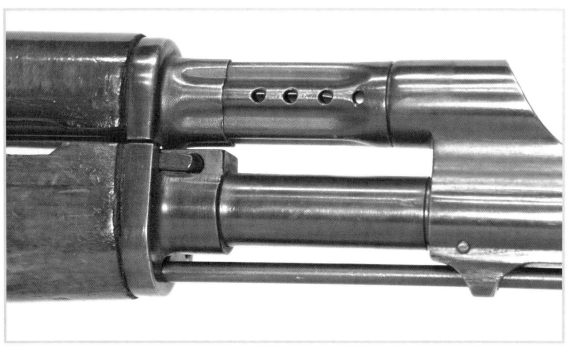

AK47 rear sight detail.

AKS47 under folding stock detail.

AKM: Kalashnikov Assault Rifle Perfection and the Worldwide Norm

The AKM was the product of Soviet Army experience, developments in Soviet industry and the continued design input from Kalashnikov and his team. This model, sometimes referred to as the AK47 Type 4, is the perfection of the AK47 design. It is this model above all that has cemented the legend of the AK and is the most widely distributed, sold and copied variant of all AK designs. It is without doubt the standard by which all other assaults rifles are judged and is still a highly credible weapon today.

Two versions were developed, produced and deployed from 1959 onwards, AKM – 'Avomat kalashnikova Modernizirovannyy', and AKMS – '*Avomat Kalashnikova Modernizirovannyy s sklady-vayushchimsya prikladom*'. In rough translation 'AKM' means 'Automatic Rifle Kalashnikov Modernised'; this is the fixed stock version and the AKMS is the version fitted with the under folding metal stock designed for airborne forces and troops working in confined or restricted spaces.

The AKM become the Soviet Army's standard assault rifle for all troops from 1959 onwards until the adoption of the AK74 in the early 1970s in its unique 5.45mm calibre. The improvements made to the AKM from the basic AK47 design were considerable and without its introduction and the production methods associated with it the spread and success of the AK may not have been what it is. Retaining the basic internal 'Kalashnikov' mechanism and layout of the AK47, the AKM introduced a wide number of changes:

- Stamped receiver construction (major change to the production method)
- Introduction of a retarder
- Improved bolt locking system
- 1,000-metre rear sight
- Muzzle compensator
- Modified fore end incorporating handgrips
- Laminated wooden stock and fore end
- Plastic pistol grip
- Rib reinforced stamped receiver cover
- Parkerized bolt and bolt carrier
- Modified front sight post
- Small 'dimple' indent either side of the receiver above the magazine well

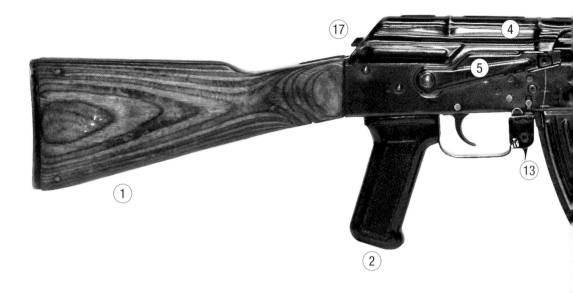

- Replacement of the 4 gas vent holes on either side of the gas tube to a position around the gas camber/gas tube join.
- Plastic 30-round magazine (though metal magazines are common on the AKM)

The first and most important change associated with the AKM was the method of manufacture. Moving away from the milled steel construction of the Type 3 AK47 the AKM was produced from stamped steel which reduced production time and waste and in addition reduced the weight of the weapon by about 1.5kg over the original AK47 Type 1. This weight reduction changed the AK from a fairly heavy weapon by today's standards into one whose weight has become the norm. In fact many modern small arms designers struggle to match the weight of the AKM even using modern materials such as Polymer and twenty-first century production methods. Weighing in at 3.3kg (and this figure includes a standard AKM plastic 30-round unloaded magazine) none of its contemporaries could match it. For example the widely used FN FAL weighs in at over 4.3kg.

Weight makes a difference to a soldier; add to the weight of the rifle itself the weight of a fully loaded magazine and the true combat weight is arrived at. For the AKM this figure with a full 30 rounds is 3.8kg. Carrying this day

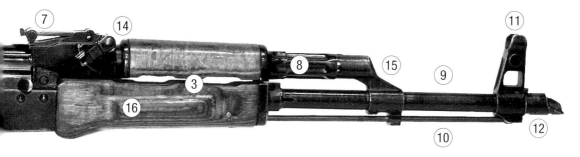

Soviet AKM 7.62 x 39mm assault rifle.

1. Stock
2. Plastic pistol grip
3. Fore arm
4. Receiver cover
5. Selector lever
6. 30-round magazine
7. Rear sight
8. Gas tube
9. Barrel
10. Cleaning rod
11. Front sight
12. Compensator
13. Magazine release catch
14. Gas tube take down lever
15. Gas block
16. Forearm finger/ hand grips
17. Recoil mechanism lug

in day out takes its toll on the soldier so getting the balance right between an effective robust weapon and 'lightness' is a fine art.

Introducing the retarder and improving the bolt locking system improved both accuracy and a steady automatic rate of fire of about 600 rounds per minute. However like all rifles firing in the fully automatic mode 'rate per minute' statistics can change by plus or minus 50 rounds or more depending on a wide range of internal and external factors. Parkerizing the bolt and bolt carrier reduced the problem of rusting and reduced 'shine' from the weapon. The cosmetic changes such as the laminated wooded stock and forearm however were only partly successful; many non-Soviet versions changed this to more hard wearing woods or plastics. It is not uncommon to see Soviet-built AKMs that are working well with the stocks in a shocking state and it's fair to say the laminated wood design is not the most hard-wearing feature of the weapon.

The introduction of the muzzle compensator however was a major change and many confuse this with a 'muzzle break'. The design aim of the compensator is to keep the barrel down when firing in the fully automatic mode. Without the compensator fitted the weapon tends to climb up and to the right. The AKM compensator is 'off set' at an angle to the left allowing the power of escaping gas following on from the fired round leaving the

barrel to push down on the metal surface of the compensator to the left and therefore compensating for the natural rise in the barrel to the right.

The Soviet design team led the away in compensator design and ultimately produced the world's best design with the compensator produced for the AK74. In addition the team also led the way in developing plastic magazines and the AKM was assault rifle to have one developed for it and successfully placed in service. Metal magazines were also produced and in most cases were favoured by non-Soviet armies. However, it is interesting to note that the current kings of AK production the Bulgarians produce excellent plastic magazines for their range of AK rifles and these are sold worldwide and are highly sought after.

Bear in mind that the AKM was designed in 1959! Kalashnikov produced a weapon chambered for the hard hitting original 7.62 x 39mm round within a light weight receiver and capable of mass cheap production. Kalashnikov and his engineering team produced the ultimate AK – the AKM and it is this model which is the real incarnation of the AK47 legend and one of the best assault rifles ever made.

THE BASICS: AK47, AKM AND AK74

AKMS 7.62 x 39mm assault rifle with under folding metal stock.

Technical Statistics: AKM
Calibre   7.62 x 39mm
Weight (with empty magazine)   3.3kg
Weight (with loaded magazine)   3.8kg
Length   880mm
Length of barrel   415mm
Number of grooves in barrel   4
Rate of fire in automatic mode   600 rounds per minute
Muzzle velocity   715 metres per second

AK74: History's Victim

In many ways the AKM was the crowning glory of Kalashnikov and his design team. However the Soviet Army continued to look for improvements though wishing to retain the reliability and simple use of the standard Soviet Army AKM assault rifle. As they had done throughout the 1940s and 1950s the Soviets looked at new and emerging weapon technologies and the result was the development of the 5.45 x 39mm M74 round.

With the US Army's adoption of the 5.56 x 45mm round in the M16 in the 1960s the Soviet Army saw the potential benefits of a smaller round with its characteristic lighter weight and higher velocity. The infantry soldier would be able to carry more ammunition into battle as a standard load. In addition the smaller round would be more accurate due to less 'kick' when fired in the fully automatic mode. Range could be increased and the higher muzzle velocity would be equally lethal on the battlefield.

The result of this thinking and the development of the 5.45 x 39mm M74 round was to produce an AK rifle capable of taking this round of ammunition. By marrying the two the Soviet Army could have all of the strengths of the AKM and the benefits of the smaller round. Kalashnikov and his team set to work and the end product was the AK74. Development and trials were conducted throughout the early 1970s the weapon entered service in 1974.

Outwardly the AK74 appears very similar to the AKM with all operating characteristics the same – however several important design changes mark out the AK74 from the AKM.

- Chambered for the M74 5.45 x 39mm round
- Rubber pad on end of stock to reduce 'kick' when fired (dropped on later models)

THE BASICS: AK47, AKM AND AK74

Late model AK74 featuring plastic furniture in plum brown colour. Note groove in the stock to aid recognition of the model and calibre (5.45mm). The pattern of the forearm retains the finger/hand 'grip' of the AKM's forearm with the addition of extra grooves to aid heat distribution and hold.

- New muzzle compensator designed to Reduce muzzle climb when fired on fully automatic and to dissipate muzzle blast
- Horizontal groove cut into stock acts as quick identifier of the weapon firing a 5.45 x 39mm round as opposed to 7.62 x 39mm ammunition

The weapon has gone through several design changes and improvements since its inception and its development history can be roughly broken down into four phases.

- Early production 1974–1977: Laminated wooden stock and forearm, thick rubber pad on end of stock (butt plate), Bakelite pistol grip, ribbed receiver top cover (like the AKM) and 45 degree gas block.
- Mid production 1977–1984: Laminated wooden stock and forearm, no rubber pad on end of stock, zig-zag pattern muzzle compensator and 90 degree gas block.
- Late production 1984–1990: Laminated wooden furniture replaced with plastic 'Plum' coloured stock, pistol grip and forearm. Smooth receiver cover.
- Current production: Black plastic furniture and black plastic magazine. Produced with both fixed and side folding stocks .Note the side folders

AKS74 showing the skeleton folding metal stock folded to the left hand side of the weapon.
1. Stock retaining latch
2. Rear sling attachment point/swivel
3. Forearm with finger/hand grips

retain a 'full' stock not a skeleton type stock of the earlier AKS-74 type. This version is still in production and available.

The muzzle compensator proved to be a masterwork of small arms design thinking. Firing the weapon on fully automatic is very controllable and therefore the Soviet Army's use of automatic fire in 3–5 round bursts as the primary mode of fire is fully matched to the design.

The compensator itself has gone through at least seven different design versions over the years. Most notably earlier models featured a zig-zig pattern cut into the side openings at the front of the compensator. The thinking on this was to break up the gases following the round out of the end of the barrel/compensator in order to reduce 'blast' sideways and instead spread it in several directions. This zig-zig cut out worked, but was abandoned in later models to reduce production time and cost. While a good feature, the weapon still functions as well without it. The main benefit is to the firer and anyone nearby!

AK74 compensator, barrel and front sight post.
1. Lateral adjustment slider
2. Detent to remove compensator

Early production models featured stocks and forearms made in laminated wood similar in design to the AKM. Over time these wooden pieces were replaced with plum coloured and then black coloured plastic stocks and forearms in addition to plastic pistol grips. The design of the forearms also changed. The original AK74 model had a forearm in the same pattern as the AKM with handgrips to aid the firer's hold. Plastic-based forearms still featured these, but the design incorporated heat-dissipating grooves to reduce heat build up on the user's hand when firing in the fully automatic mode.

Then most notable change to the AK74 over time was the change in the gas block/chamber from its original 45 degree angle – as found on the AKM – to a 90 degree angle developed and produced on AK74s from about 1977 onwards. Accuracy and effective range of the AK74 compared to the AKM is greatly improved. However, there is debate over whether a smaller higher velocity is as deadly as a larger round. The author's view on this subject is that a larger round would have been more effective and Kalashnikov himself said that the 7.62mm round should have been further developed and incorporated into the improvements made in the AK74. Had the Soviets produced a 6mm round for the AK74 it would have been perfection.

Despite the debate over calibre the AK74 was a success and it proved itself at least if not more reliable then the superb AKM. Its problem however was history and the monumental charge in the world order with the end of the Soviet Union and the governments of its eastern block allies.

The AK74 became the Soviet Union's main frontline assault rifle, seeing heavy fighting in Afghanistan. As with the previous AK47 and AKM rifles and their associated 7.62 x 39mm ammunition it was expected that the wider Warsaw Pact allies would in time follow suit and adopt the new 5.45 x 39mm M74 round, maintaining the commonality of ammunition and small arms that had been a feature of the Warsaw Pact weapons program. East Germany (DDR) and Poland began the process but then the Soviet Union began to collapse and the armies of Eastern Europe were still largely armed with 7.62mm weapons when the Berlin Wall came down. Whereas the AKM had become the norm of the socialist world, the AK74 only reached a few; Russia and its then CIS counties were the only bulk users.

This situation remains the same at the time of writing. Russia retains the 5.45 x 45mm round as its primary assault rifle calibre mainly in the AK74 and in the new Russian Army AN-94 rifle. In addition the former countries of the CIS with their stocks of former Soviet Army AK74s retain them as their principle weapon.

No country except North Korea (DPRK) has adopted the 5.45 x 39mm round since the end of the Soviet Union, however the design improvements of the AK74 are alive and well most notably in the excellent Russian Kalashnikov AK103. This weapon retains the design of the AK74 and its excellent muzzle compensator, and is produced in the original 7.62 x 39mm calibre – perhaps it is this design that Kalashnikov would have wanted the Soviet Army to adopt in the early 1970s rather then following the small calibre trend.

The author has handled and fired the AK74 on a number of occasions – it's a great rifle to shoot partially in the fully automatic mode, totally reliable and more then accurate enough for an infantry battle. Had the Soviet Union lived on the AK74 would have been produced and distributed in the same way as its older brother the AKM but the march of history put an end to that.

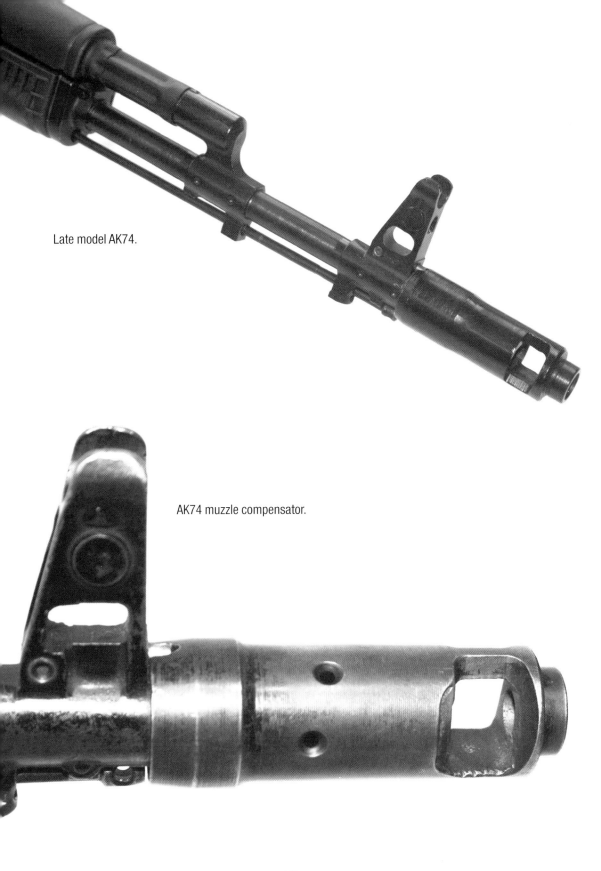

Late model AK74.

AK74 muzzle compensator.

AK47 ASSAULT RIFLE

AKS74

Technical Statistics: AKS74

Calibre	5.45 x 39mm
Weight	(with empty magazine) 3.3kg
Weight	(with loaded magazine) 3.6kg
Length	940mm
Length of barrel	415mm
Number of grooves	4
Rate of fire in automatic mode	600 rounds per minute
Muzzle velocity	900 metres per second

AK74 sling.

AKS74

Bulgarian AK74.

CHAPTER THREE

INSIDE THE AK47

What makes the AK and more specifically the AKM so good is its inner mechanisms. The weapon is a gas operated rifle with a rotating bolt fed from a 30-round magazine and fires from a closed bolt position. The gas operation used with the AK design is known as a long stoke system, which means the gas piston which is attached to the bolt carrier runs the full length from the breach area to the gas chamber via the gas tube sitting over the barrel. Essentially it is just one long piece of steel, allowing little room for malfunction. It's this one piece of metal that moves backwards and forwards under the pressure of gases diverted from the barrel to the gas chamber and onto the face of the piston. This pressure pushes the whole bolt carrier (which contains the bolt with the firing pin) backwards which in turn is pushed forward again under the pressure of the recoil mechanism (spring) from the rear of the weapon. Each time the trigger is pulled the hammer is released and it strikes the rear of the round of ammunition contained in the breach. The bullet flies down the barrel followed by gases (produced from the cartridge being struck by the hammer) some of which are directed via a small hole in the barrel up a 45 degree tube cut into the gas block and into the gas chamber and on the face of the gas piston.

When the gas piston moves backwards the spent cartridge contained in the breach is ejected by a small piece of metal on the bolt called the extractor. The spent case is hooked and flipped out the right side of the weapon in the space created by the bolt carrier moving backwards. As this is happening a new round of ammunition moves upwards under the pressure of

AK47 ASSAULT RIFLE

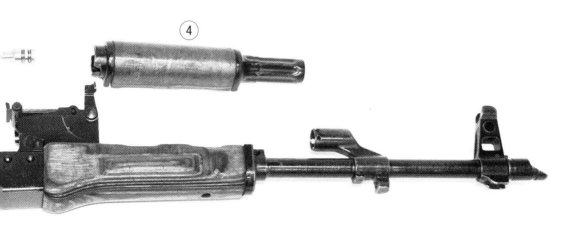
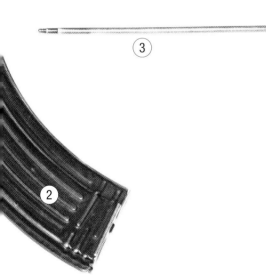

AKM disassembled for cleaning.
1. Stock, receiver and barrel assembly
2. Magazine
3. Cleaning rod
4. Gas tube
5. Bolt
6. Bolt carrier and gas piston
7. Recoil mechanism
8. Receiver cover

AK47 ASSAULT RIFLE

Soviet AKM selector lever in upper safe position.

AKM gas tube and forearm detail. Note gas tube take down lever.

the spring in the bottom of the magazine and as it does so gets pushed and drilled into the breach by the now rotating bolt returning under the pressure of the recoil mechanism. All this happens in a split second and so long as you have rounds in the magazine and you charge (cock) the weapon and pull the trigger it will keep doing it.

Space is a key factor in the success of the weapon and within the rear of the receiver the space between the recoil mechanism and the internal base of the weapon is great. In addition the small space between moving parts like the bolt carrier and the receiver wall all allow for dirt, sand and water to have little effect in jamming up the system.

- Only seven moving parts
- Long stroke gas piston system
- Lots of space in the rear of the receiver (body of the rifle)
- Chromed barrel
- Two lug bolt
- Calibre
- Lightweight

The two features of the space between moving parts and the chromed barrel go some way to explain why the weapon has been so successful in two major conflicts. The chroming of the barrel prevents rusting even if the weapon is not regularly cleaned, ideal for the wet terrain of Vietnam. Space between parts gives room for sand to fall though and not stop the functioning of the system. Given its use in recent years in the desert like condition of Iraq and Afghanistan its no surprise the AK is the weapon of choice for many a fighter.

Calibre is also key – the 7.62 x 39mm round is a killer – it was designed as such with no pretence or notion about wounding or reducing soldier's weight in combat. If you're faced with an enemy who wants to kill you the last thing you need is a rifle that fires an inadequate round – it has got to knock him down first time and keep him down. The 7.62 x 39mm does that and combined with the simplicity of the AK design the complete package is hard to beat.

AKM forearm.

AKM barrel and front sight post.

AKM barrel, front sight post and compensator.
1. Compensator
2. Front sight lateral adjustment slider
3. Detent (for removal of compensator)
4. Front sight guard
5. Cleaning rod
6. Gas block

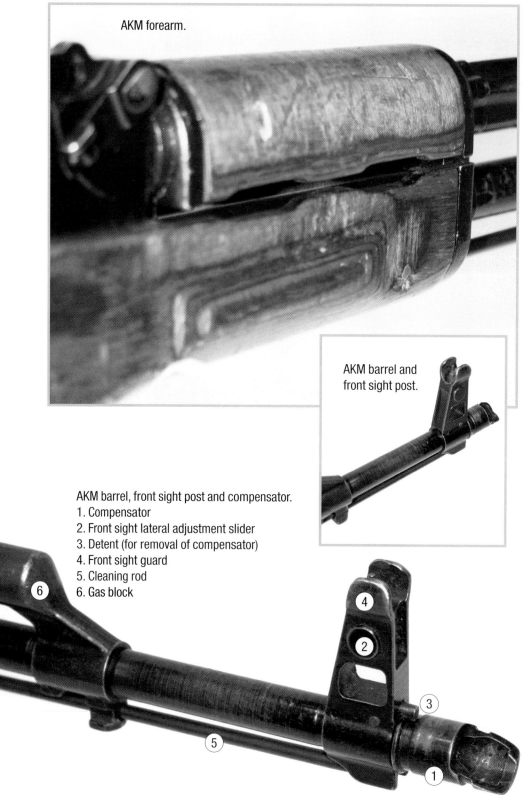

Underside of AKM gas tube.

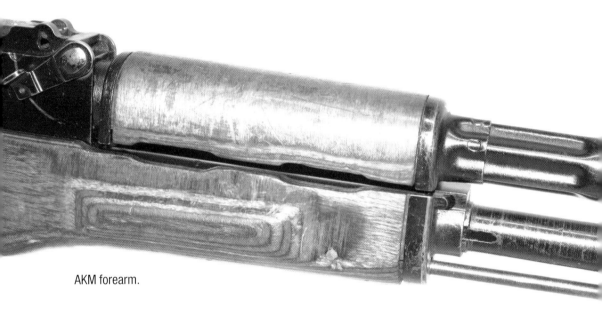

AKM forearm.

AKM magazine lips and follower.

AKM with gas tube removed showing gas block detail.

AKM with receiver cover removed – note recoil mechanism fitted into rear of bolt carrier.

INSIDE THE AK47

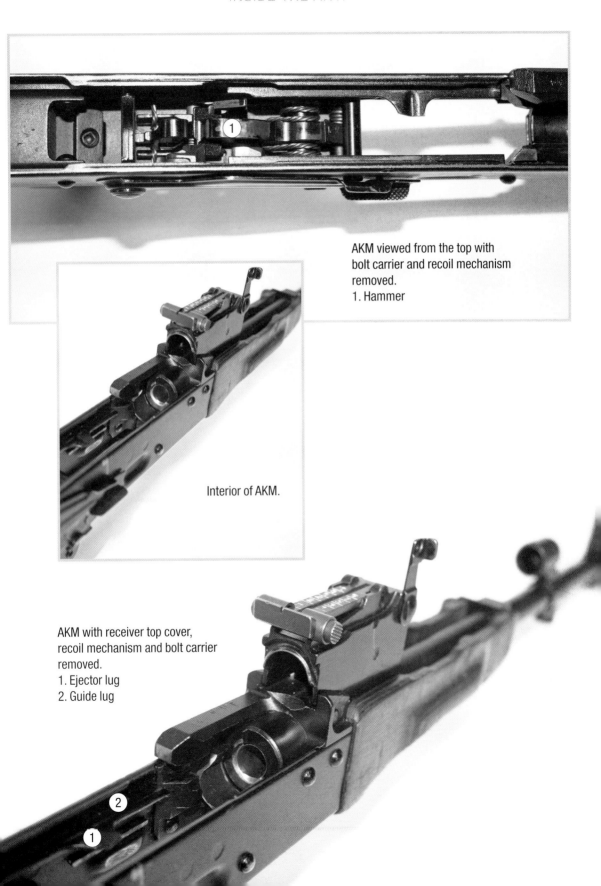

AKM viewed from the top with bolt carrier and recoil mechanism removed.
1. Hammer

Interior of AKM.

AKM with receiver top cover, recoil mechanism and bolt carrier removed.
1. Ejector lug
2. Guide lug

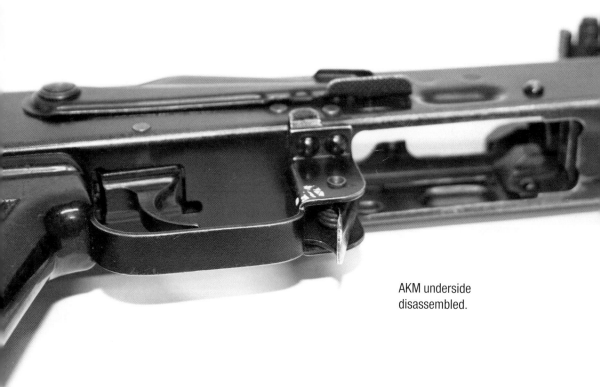

AKM underside disassembled.

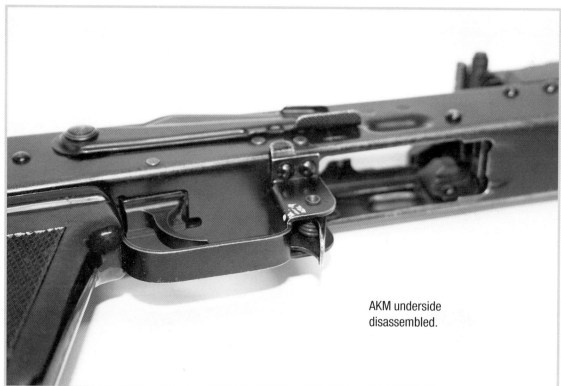

AKM underside disassembled.

AKM with receiver cover removed showing position of recoil mechanism. Note the space created between the recoil mechanism and the body of the rifle. This use of space is a key factor in the success of the design, allowing dirt, sand and water to fall freely through without fouling the working parts and causing stoppages.

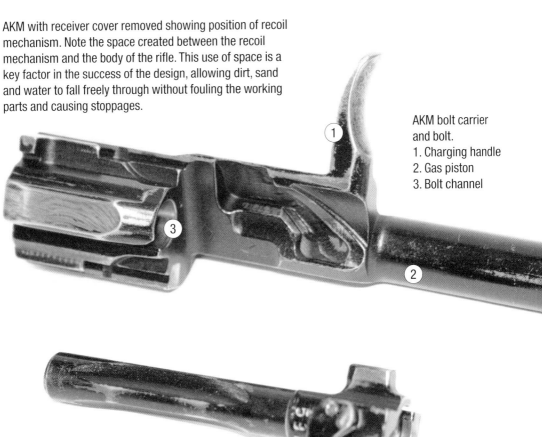

AKM bolt carrier and bolt.
1. Charging handle
2. Gas piston
3. Bolt channel

AK47 ASSAULT RIFLE

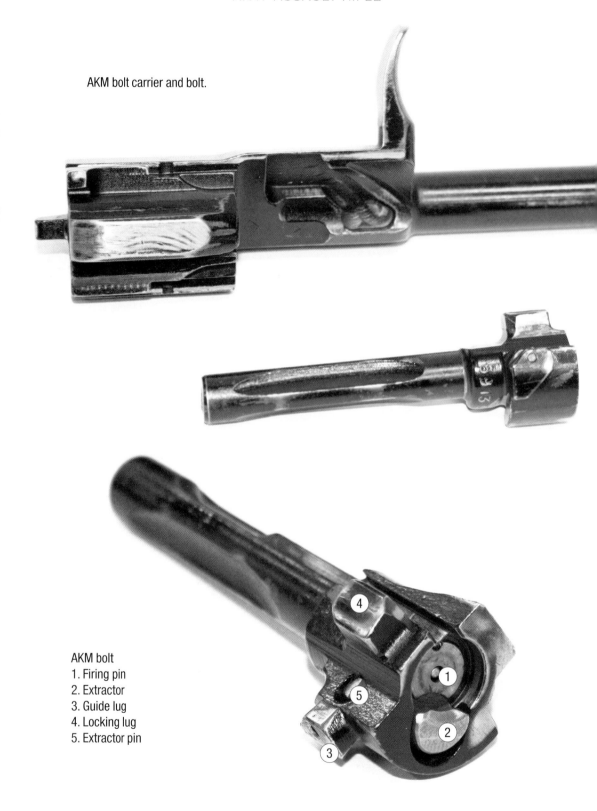

AKM bolt carrier and bolt.

AKM bolt
1. Firing pin
2. Extractor
3. Guide lug
4. Locking lug
5. Extractor pin

AKMS metal under folding stock detail.

AKM cleaning rod.

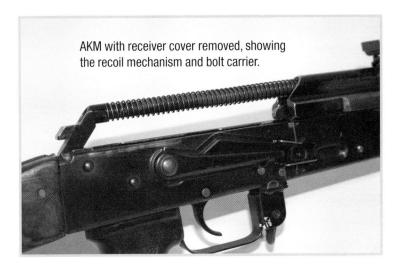

AKM with receiver cover removed, showing the recoil mechanism and bolt carrier.

AKM top view.
1. Receiver cover – note ribbed reinforced design.
2. Recoil mechanism lug – pressed in to remove the receiver cover and the recoil mechanism from the body of the receiver.

INSIDE THE AK47

AKM – note 'dimple' above the magazine on the body of the receiver which acts as a guide in low light conditions for the magazine and helps retain it in position.

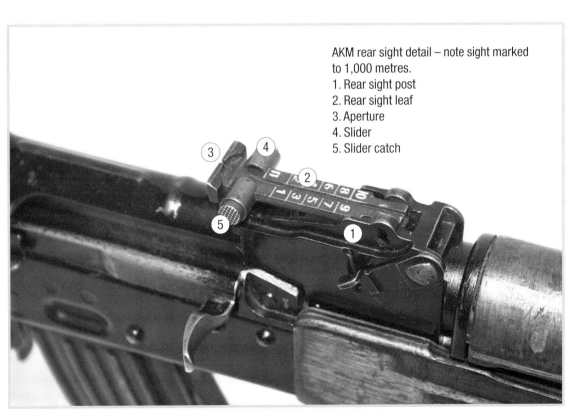

AKM rear sight detail – note sight marked to 1,000 metres.
1. Rear sight post
2. Rear sight leaf
3. Aperture
4. Slider
5. Slider catch

AKM rear sight.

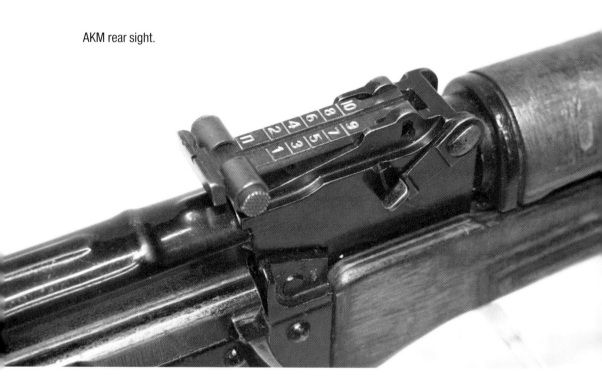

CHAPTER FOUR

HOW TO USE IT

Operating the AK47 is the easiest way to demonstrate the simplicity and robustness which have made it such a successful weapon. The author conducted an experiment in 2009 to see how quickly he could train a person to use an AK. Taking a man who had no previous experience of handling any military firearm the aim was to enable him to

- Carry out normal firearms safety
- Load the weapon
- Set sights and fire
- Unload the weapon

The total time it took to complete the experiment was an amazing 4.5 minutes! This was of course a super fast test but it shows how easy it is to turn a civilian into a gunman. This explains in part why so many semi-trained militias around the world arm themselves with the AK.

The experiment was conducted using a Hungarian-built version of the AKM in dry classroom conditions and the individual being trained was an educated middle class city worker. Conditions in the real world are of course very different and militia armies and conscript units will contain a mixture of human subjects ranging from the well educated to those with little or no formal education. The AK was designed for the Soviet Army whose soldiers spoke a variety of languages and were not equally educated. Furthermore, they were expected to fight in a wide spectrum of climates. It's therefore no wonder the AK can fit into any condition around the globe and that anyone can use it.

However, in order to properly train an individual in the weapon and give correct training in disassembly, assembly, cleaning, magazine loading and correct use of sights would take longer – however this process can be achieved in approximately four hours. In order to do this the process can be broken down into eight lessons prior to getting onto the firing range. Each lesson is conducted with an instructor demonstrating and the trainee repeating each stage and repeating and repeating until learnt. This process is often referred to as training the muscle memory so as soon as the new soldier picks up his or her AK the use of it is retained in muscle memory, be it the next day or in months to come in the case of reserve troops on refresher training.

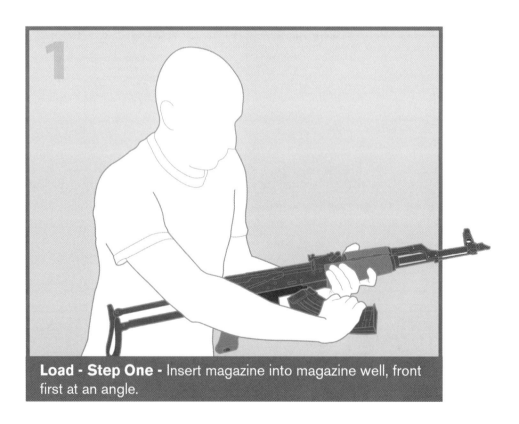

Load - Step One - Insert magazine into magazine well, front first at an angle.

HOW TO USE IT

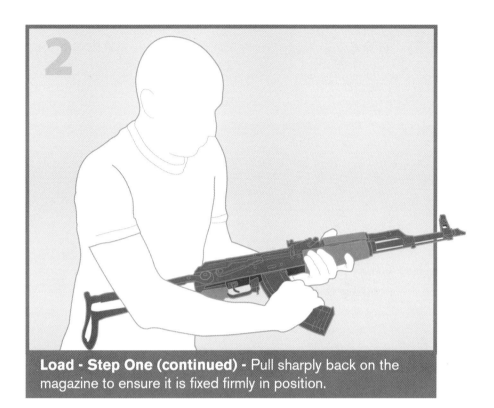

Load - Step One (continued) - Pull sharply back on the magazine to ensure it is fixed firmly in position.

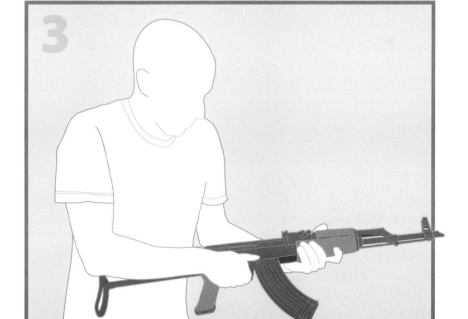

Load - Step Two - Push down on the selector lever to the fully automatic or semi automatic fire position.

AK47 ASSAULT RIFLE

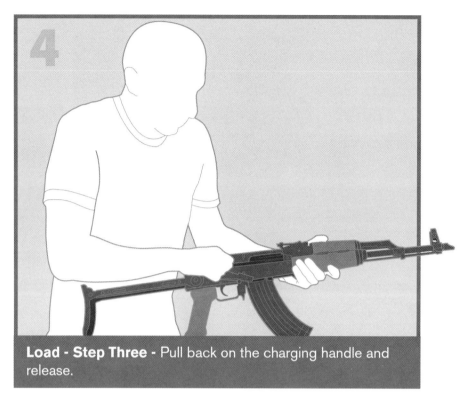

Load - Step Three - Pull back on the charging handle and release.

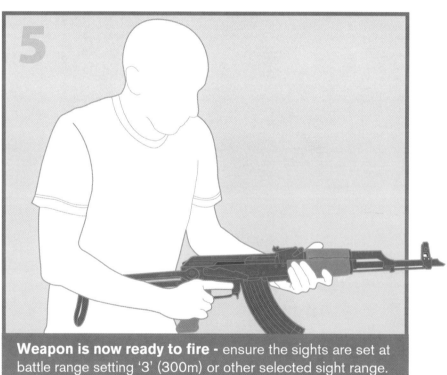

Weapon is now ready to fire - ensure the sights are set at battle range setting '3' (300m) or other selected sight range.

HOW TO USE IT

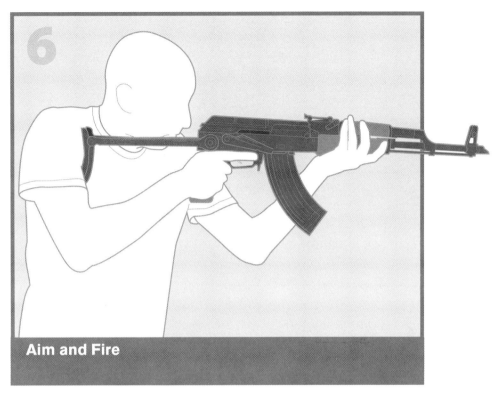

Aim and Fire

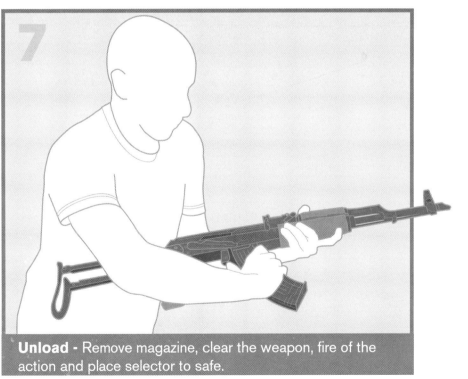

Unload - Remove magazine, clear the weapon, fire of the action and place selector to safe.

Lesson One: Characteristics of the Weapon

To include calibre, weight, magazine capacity, firing modes (safe, semi automatic and fully automatic) main external features such as position of sights, stock, barrel, forearm, pistol grip, selector lever ,magazine, magazine well and trigger.

Lesson Two: Weapon Safety

Pick up the weapon, remove the magazine, depress the selector lever down from the safe position to the fully automatic position, pull back the charging handle, inspect inside the weapon via the ejection opening to ensure no live rounds are present. Release the charging handle, fire off the action (by pressing the trigger) and re-set the selector to the upper safe position.

Lesson Three: The Load

Holding the forearm with the left hand and a magazine in the right hand (at about a 30 degree angle) slot the magazine into the magazine well and pull back until the magazine snaps home. Note on the front of the magazine is a metal 'lip' this needs to sit inside the front of the magazine well when slotting it into place.

Once the magazine is in place and using the right hand, depress the selector lever down from the safe position to either the fully or semi-automatic position (note the first 'click' down is fully automatic and the last position is semi automatic). Then using the right hand pull fully back on the charging handle and release. The weapon is now loaded and made ready to fire. Place the right hand onto the pistol grip and the left hand should remain on the front forearm.

Lesson Four: The Unload

Remove the magazine using the right hand, grasp the magazine and using the thumb (at the same time) depress the magazine release catch forward while pulling forward the magazine until it is released from the magazine well. Pull back on the charging handle (if the selector lever is the upper safe position at this stage the lever needs to be depressed down to the fully automatic position) and check to see no rounds remain in the body of the weapon. Release the charging handle, fire off the action by pressing the trigger and replace the selector lever into the upper safe position. The weapon is now fully unloaded and safe.

Lesson Five: Sight Setting

The rear sight determines the range setting on all AK models. AK47s are sighted to 800 metres and AKMs to 1,000 metres. By simply pinching on either side of the rear sight slider you can move the sight up and down the range scale depending on the range required. To make things simple the very rear position is a 'battle' setting and will sight the weapon up to 300 metres without having to change sight positions. Given that most infantry combat happens at less then this range, and allowing for the characteristics of the weapon, keeping the weapon set to this position or the 300 metre mark is suitable for battlefield use.

Note for zeroing the weapon the front sight post is used to correct the rifle for both elevation and lateral sight adjustment. Zeroing and the correction required for the front sight post is a skill and requires further training and practice outside the needs of basic training – it needs to be taught on the firing range and requires specific tools.

Lesson Six: Magazine Filling

Holding the magazine in the left hand with the curve of the magazine body towards the palm of the hand feed the individual rounds of ammunition into the body of the magazine by pushing down and backwards. Each magazine will hold 30 rounds. Inspect the 'lips' of the magazine to ensure they are not damaged prior the loading and ensure each round is correctly seated into the magazine.

Lesson Seven: Stoppages

Clearing a stoppage with an AK is a simple process; pull back on the charging handle and continue firing! If the stoppages continue, field strip the weapon, clean, reassemble and continue. Stoppages on AKs can occur due to the following problems:

- Faulty ammunition
- Worn out parts
- Excessive fouling
- Damaged magazine

If the weapon is submerged in water or sand remove the magazine, pull back on the charging handle and shake out the sand or water then reload and continue. Not much stops an AK from firing and in the author's experience it is mainly down to poor or old ammunition rather then the weapon itself. In all cases the author merely re-charged the weapon and continued firing.

Lesson Eight: Disassembly, Assembly and Cleaning

This is where the AK shines through. It's a simple process and once the weapon is disassembled all of the working parts and the inside body work (receiver) can be got at in order to clean it. It's not fiddly and has no small parts that can be lost or broken in the process. Ensure the weapon is not loaded by carrying out the normal safety check.

Step One	Remove magazine
Step Two	Remove the combination tool from its position within the stock (fixed stock model only)
Step Three	Remove the cleaning rod
Step Four	Remove the receiver cover
Step Five	Remove the recoil mechanism from the receiver
Step Six	Remove the bolt carrier (containing the bolt)
Step Seven	Remove the bolt from the bolt carrier
Step Eight	Remove the upper forearm and gas tube

The weapon is now field stripped for cleaning and inspection. In addition to the above disassembly process the user can also remove the forearm and the compensator if one is fitted. By removing the compensator the process of cleaning the barrel using the cleaning rod and elements of the cleaning kit contained within the combination tool is made easier and is indeed the correct procedure.

The time it takes to disassemble the weapon is remarkably short and can be done in 30 seconds with practice. However, in order to lessen wear on the parts it is recommended to slow this down. When disassembling the AK set out the parts on a clean surface in the order of disassembly as an aid for reassembly. To assemble the weapon repeat the above in reverse order.

Each of the first seven lessons can be learnt in twenty minutes each and the final lesson in disassembly, assembly and cleaning takes 40 minutes. The remaining time of an hour is taken up in testing the user in each stage until 100 per cent correct. This method gives a total training time of four hours, at the end of which the pupil will be expert in the use and maintenance of the AK and ready for the range and combat training as a competent user.

This quick time approach to weapon training allows more time to be dedicated to tactical training and time on the range. During the Second World

War the Soviet Army had little time to train infantry soldiers and generally only a few weeks was available. This experience led to the need for an effective weapon which could be learnt in a few hours and the soldier sent on his way to battle. The Vietnamese communist forces in South Vietnam were similarly pressed for training time and had few trained instructors, problems which were overcome by the AK's simple construction and ease of use. In addition reserve or part time troops – such as the various 'workers militias' of the former Communist eastern block – could be trained in a matter of hours, and given refresher training in the evenings and at weekends with no requirement for the soldiers to re-learn complicated systems or the need for highly skilled instructors. In modern times the story is the same – 'militiamen' in the Middle East or Africa can be trained with ease and from a young age to handle the weapon and still have a 'day job'. The majority of Hezbollah's fighters in Southern Lebanon and their allies in the Lebanese Communist Party that faced the Israeli invasion in 2006 were part time fighters working in their communities and training in their own time.

The AK is at its best firing on fully automatic at ranges below 200 metres. The weapon was designed in the first place to fire on automatic in contrast to western small arms designs that are designed to fire single aimed shots and have a fully automatic mode for use at very close range such as in rooms and trench clearing. Firing in bursts of 3–5 rounds is the weapon's main operating condition and its primary firing method. Many western small arms observers have tried to compare the AK with western designs such as the M16 but this is a false comparison. The M16 is first and foremost a true rifle designed to fire in the semi automatic mode with aimed accurate shots. The AK on the other hand is an assault rifle born out of the need for a massive volume of firepower in the automatic mode. The AK should always be judged in this way, fired in bursts of 3–5 rounds as part of a co-ordinated assault or defensive plan.

Zeroing the AK

In order to gain the best possible accuracy from the weapon and indeed any combat rifle the AK should be 'zeroed' by bringing the rifle to a state of 'normal zero'. The process is fairly simply but does require the use of two tools. First the screwdriver blade from the combination tool kit, and second the windage adjustment tool – this is a separate piece of equipment that was issued at company level within the former Soviet Army and is a rare item. This adjustment tool resembles a small 'G clamp' with a wined-in bolt that is pressed against the slider and when tightened by screwing in the bolt moves the slider without damaging the front sight post. The screwdriver blade contains at its base an insert that is used to screw in or out the front sight blade whereas the windage adjustment tool works on the lateral windage slider found on the side of the front sight post.

Zeroing should take place on a range in good weather conditions at a range of 100 metres. Four shots should be fired at a control point and then the front sight adjusted to up or down, left or right until the 'mean point of impact' is on this control point. To calculate and make the adjustments the following guide should be used (based on a Soviet AKM)

- If the mean point of impact is below the control point, screw the front sight blade in
- If the mean point of impact is above the control point, screw the front sight blade out
- If the mean point of impact is to the left of the control point, move the slider to the left
- If the mean point of impact is to the right of the control point, move the slider to the right

Making the adjustments to both the slider and the sight post is a 'hit and miss' affair and takes practice to get right in a short time period. A 1mm movement of the slider either way will move the mean point of impact about 25cm at 100 metres and one complete turn of the fight sight blade will move it by 20cm at 100 metres.

In theory all AKs are issued with a cleaning kit/combination tool containing a screw driver blade so making adjustments up or down on the front sight blade is straight forward. The availability of the correct issued lateral adjustment tool for the slider however is another matter. In the absence of this rare item a small hammer and suitable small metal chisel type object will work.

Fire the four round groups again after each small adjustment until zeroing is complete and this should normally take 2–5 groups to get right. Once done the need to re-zero is infrequent and given that the weapon is primarily designed to operate in the fully automatic mode as part of a tactical grouping in defence or attack its accuracy is good enough.

CHAPTER FIVE

WORLDWIDE

The AK is now a worldwide 'brand' with production in its original AK47, AKM, AK74 or design influence variations having taken place in some nineteen countries. The states of Eastern Europe under communist leadership were the first to take the weapon into production with the Warsaw Pact looking to standardise weapon design and most importantly ammunition calibres around the Soviet models.

Today production and export are based in China, Iran, Algeria, Russia, Romania, Bulgaria, Sudan, Cuba and North Korea. This listing does not include countries producing assault rifles where the design is largely or heavily influenced by Kalashnikov's AK work. China is the biggest seller, but Bulgaria is continually finding new markets for its high quality milled receiver AK47 copies.

The weapon is also made illicitly within countries that do not produce the weapon though a recognised arms manufacturing company. These include breakaway regions of Burma and the tribal areas of Pakistan close to the Afghan broader where AK production, repair and rebuild has gone on for over 30 years.

Below is a summary of the countries and variants based purely on the original three Soviet designs of the AK47, AKM and the AK74 in the 7.62 x 39mm and 5.45 x 39mm calibres. In the interest of time and space, versions produced in the NATO 5.56 x 45mm calibre have not been covered. In addition only 'assault rifle' versions are included, not sub machine-gun versions of the full assault rifle or specialist grenade launching sub-variants.

Civilian patterns produced in semi–automatic only modes such as the US/Romanian WASR range are also outside the scope of this chapter.

Algeria

Algeria produces two versions of the AK, one with a fixed wooden stock and the second with a metal under folding stock. Both are in the 7.62 x 39mm calibre and appear to be similar in design to the Chinese Type 56-1 rifle.

Model 89
7.62 x 39mm assault rifle with fixed stock. Close resemblance to Chinese Type 56 series excluding the fixed spike bayonet.

Model 89-1
7.62 x 39mm under folder with close resemblance to the Chinese Type 56-1.

Albania

Albania's first use of the AK47 was with Soviet-imported fixed stock and folding stock Type 3 AK47s. With the country's political shift away from the Soviet Union and its focus on China as an ideological model in the 1960s, the supply ceased. Chinese Type 56 weapons supplemented the Soviet supply until local production took place. Albanian-made AKs are a rare find with few exported from the country. In 1974 the Albanian 'Gramsh' state arsenal began licensed production of Chinese Type 56 rifles with two variants produced.

Type A
Copy of the standard fixed stock with folding bayonet Chinese Type 56 and under folding Type 56-1. This is the most common Albanian model of AK.

Type B
Based on a fixed stock Type 56 but fixed with a spigot type grenade launcher on the barrel of the weapon. In addition the rear sight is repositioned to the middle of the receiver cover.

WORLDWIDE

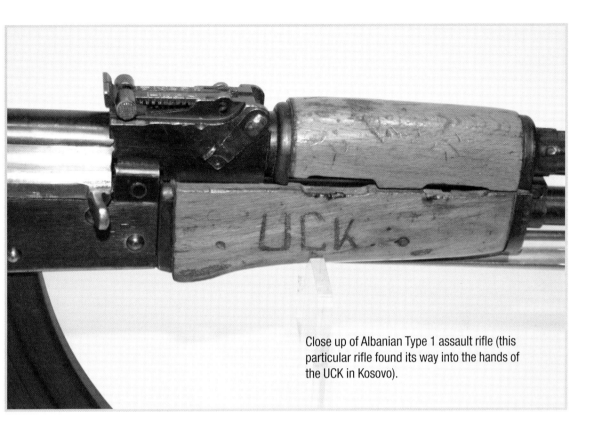

Close up of Albanian Type 1 assault rifle (this particular rifle found its way into the hands of the UCK in Kosovo).

Albanian Type 1 assault rifle front right hand view; note the four gas vent holes on the gas tube of the type found on the AK47 and smooth sided forearm furniture. It has no finger/hand grips as found on an AKM.

Burma (Myanmar)

The AK47 is manufactured in the north of the country by forces opposed to the Military junta running the country. These forces are peoples of 'Chinese' ethnic background and it appears that the weapons are copies of the Chinese Type 56 series – mainly under folders.

Bulgaria

The Bulgarians are now the kings of AK production producing the best AKs in the world. Manufacturing quality is second to none and they have largely retained the milled receivers of the later model AK47s.

In-country production began in the early 1960s with a copy of the Soviet AK47 and later the AK74 but never producing a copy of the AKM. Bulgarian AK47s feature plastic stocks, pistol grips and forearms often in a reddish-brown colour giving them a distinctive look. Current production models however are all produced in a matt black plastic furniture finish.

Over the years however, the Bulgarians have managed to incorporate some features of both the AKM and AK74 in their designs while retaining milled

Bulgarian AK47 fitted with brown plastic furniture. All other features are a direct copy of the original Soviet AK47 Type 3. In Bulgarian service this weapon is known as the AKK and AKKS (under folding stock).

receivers and developing flash hiders. In addition they have refined the AK magazine producing highly efficient plastic magazines of a unique pattern. Often referred to as 'waffle' pattern due to the raised chequered pattern on the outside of the magazines, these magazines are very well respected.

Bulgarian AKs are sold all over the world and provide buyers with a quality choice option at cheaper prices to other European and North American manufactured assault rifles.

AKK
Copy of AK47 Type 3 fitted with reddish-brown fixed stock, pistol grip and forearm. 7.62 x 39mm. Standard service rifle of the Bulgarian Armed Forces.

AKKS
Under folder metal stock version of the AKK.

AR
Copy of the AK47 with fixed stock, milled receiver and black plastic furniture. Some versions have been produced with a 90 degree angle gas block as well as the 45 degree gas block. 7.62 x 39mm calibre.

Bulgarian AK74 compensator.

AR – F
Under folder version of the AR. 7.62 x 39mm calibre.

AR – 1
Fixed stock version of AR fitted with a Bulgarian designed flash hider. 7.62 x 39mm calibre.

AR – 1F
Under folder version of the AR – 1. 7.62 x 39mm calibre.

AR – M1
Fixed stocked rifle with a milled receiver and incorporating features from the AKM and AK74 – a hybrid in 7.62 x 39mm calibre. The gas tube is of the AKM type with the gas vents arranged around the join of the tube and the gas chamber/block. The gas block itself is set to 90 degrees and has a Bulgarian-designed flash hider fitted to the barrel. Front sight post is similar in style to that found on an AK74. All furniture is black plastic. 7.62 x 39mm calibre.

Bulgarian AK74.

AR – M1F
Under folder version of the AR – M1 with a metal stock design similar to that found on the AKMS. This version like the AR – M1 is a total masterpiece incorporating the best of the AK47, AKM and AK74. 7.62 x 39mm calibre.

AR – M7F
Side folder version of the AR – M1 with a full stock folding to the left of the receiver.

AK74
Direct copy of the Soviet original design incorporating the zig-zag pattern muzzle compensator of mid-period Soviet AK74 design. Earlier models featured laminated wooden furniture in a light colour referred to as 'blonde'. Later models were produced with black plastic furniture similar to the Soviet AK74.

Chinese Type 56 assault rifle, early version based on milled receiver (note large rectangular cut-out above magazine on the receiver body). Wooden stock, pistol grip and fore end and under folding spike bayonet. 7.62 x 39mm calibre. Front sight is covered rather than open as in Soviet AK47 models. This feature is typical of Chinese manufacture and often a clear sign that a particular model is Chinese or the manufacturing process was Chinese-influenced.

Cambodia

Unlicensed copies of Soviet AKMs have been produced in-country and closely copy the original design, although different stock and forearm materials and colourings have been found. In addition some examples show four gas vent holes along the gas tube similar to the Chinese Type 56 model.

China

China has been, outside of the former Soviet Union, and is now the single biggest producer and exporter of the AK47 and its derivatives. Across the globe Chinese built weapons can be found in every level and type of formal army, police, para-military and militia organisation. Africa is awash with Chinese guns as is the Middle East and Afghanistan. If you see an AK on the news channels today there's a good chance it is Chinese built. Chinese AKs are cheap, solid and available to anyone with hard cash or Chinese economic interests to trade. Find a battlefield and your find a Chinese AK.

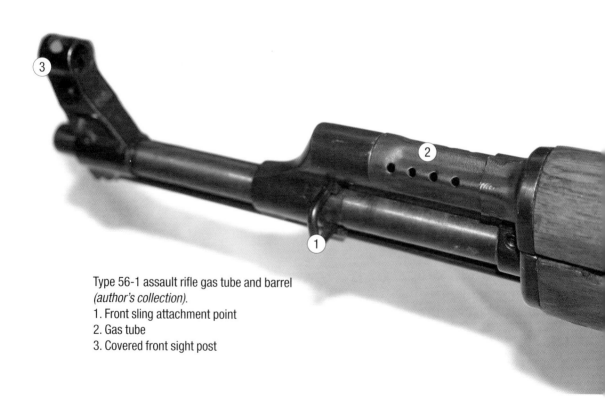

Type 56-1 assault rifle gas tube and barrel *(author's collection)*.
1. Front sling attachment point
2. Gas tube
3. Covered front sight post

Chinese Type 56-1 assault rifle. This model is fitted with the under folding metal stock as per the Soviet AK47 under folder. Note the AKM style dimple above the magazine on the receiver body in contrast to the larger rectangular cut out of the older model Type 56. The front sight post is of the covered Chinese type and the muzzle break is unique to the Chinese produced models. The weapon retains features of the earlier AK47 such as the four gas vent holes along the gas tube and a pin steel bolt carrier. Chinese AKs are often a mixture of the AK47 and later AKM models *(author's collection)*.

Type 56

Originally a direct copy of the fixed stocked Soviet AK47 complete with milled receiver and no folding spike bayonet that has become the hallmark of the later Type 56 rifles. Production of this direct copy was soon switched to the Type 56 we know today. From the late 1950s until the mid-1960s Type 56 rifles were produced with the milled receiver of the AK47 then moving to the stamped receiver of the AKM. Chinese weapons on either milled or stamped receivers are distinguished by the following features:

- Fully enclosed (hooded) front sight – this is the most recognisable feature of Chinese produced AKs
- 4 gas vents (holes) along the gas tube
- Receiver made from thicker and heavier steel – 1.5mm thick as apposed to 1mm thick on the Soviet AKM original
- Spike folding bayonet – note however some fixed stocked versions are and were produced without this item
- Smooth receiver top cover lacking the 'ribs' of the AKM

Type 56-1

Under folding stock version of the above; widely used and encountered all over the globe.

Type 56-2
Side folder version of the Type 56-1 which features the unique Chinese style curved pistol grip first introduced on some Type 56-1 rifles. This version is becoming very popular and heads the current Chinese export field for AK type rifles.

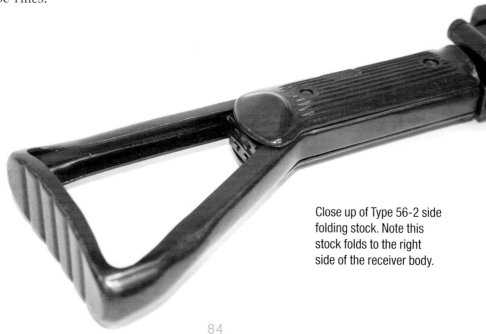

Close up of Type 56-2 side folding stock. Note this stock folds to the right side of the receiver body.

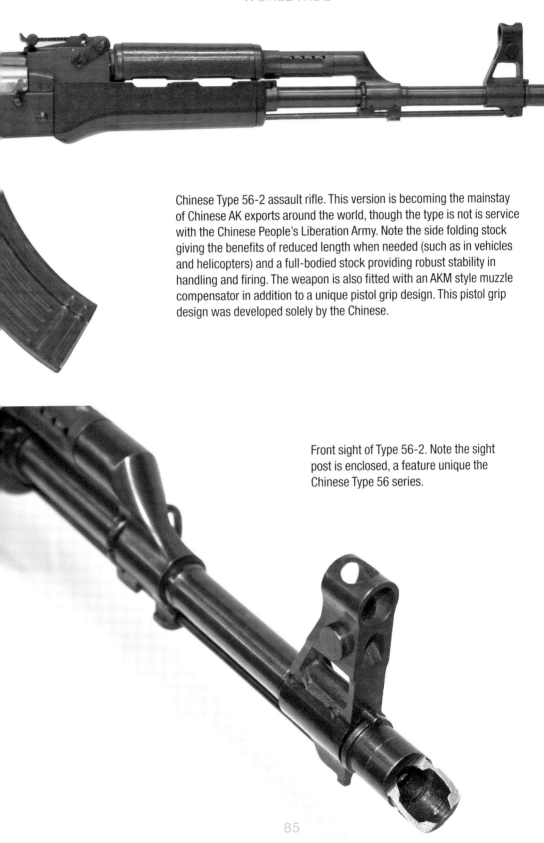

Chinese Type 56-2 assault rifle. This version is becoming the mainstay of Chinese AK exports around the world, though the type is not is service with the Chinese People's Liberation Army. Note the side folding stock giving the benefits of reduced length when needed (such as in vehicles and helicopters) and a full-bodied stock providing robust stability in handling and firing. The weapon is also fitted with an AKM style muzzle compensator in addition to a unique pistol grip design. This pistol grip design was developed solely by the Chinese.

Front sight of Type 56-2. Note the sight post is enclosed, a feature unique the Chinese Type 56 series.

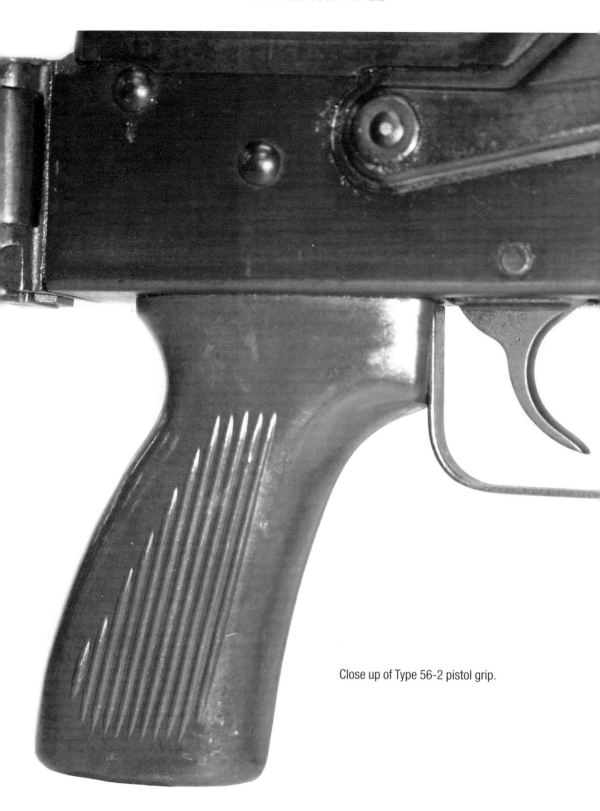

Close up of Type 56-2 pistol grip.

Cuba

Cuba produces both the AKM and AKMS. The variants are exact copies of the Soviet originals maintaining the 7.62 x 39mm calibre. Interestingly the Cubans have now developed there own optic battle sight for the weapon which is attached to the rear sight base and extends to the middle of the receiver cover in a thin rectangular box configuration. Little is know of its design although the author has seen a number of photographs of the sight and clearly the Cubans have looked to improve the battle range hit capability.

Democratic People's Republic of Korea (DPRK)

The DPRK – more commonly known as North Korea – has been a producer of the AK series since the 1950s. Mirroring the Soviet designs, the DPRK has produced indigenous versions of the AK47, AKM and AK74 with some local modifications. The weapons produced are all of a high standard and the slight changes made have improved on the original designs.

The weapons have been exported, notably to Africa where the latest DPRK variant of the AK74 has been seen. In addition small numbers were sent to North Vietnam during the Vietnam War but trade is nowhere near the level enjoyed by Chinese manufacturers.

Type 58
Near identical copy of the original Type 3 Soviet AK47, produced with both solid fixed stock and under folding metal stock. These Type 58s are still very much in service in the DPRK with units of the Workers-Peasants Red Guard parading with the weapons in Pyongyang in 2009. Given that the strength of the militia-based Guard is up to 4 million people strong gives some indication to the level of production of the weapon in the DPRK.

Type 68
Copy of Soviet AKM both fixed stock and under folding versions. The North Koreans have made some modifications to the original design notably the removal of the handgrips on the forearm and therefore retaining the smooth side finish to the forearm of the original AK47. This may have been retained to better fit the smaller hand size of the average Korean; however, the same feature was retained by the Hungarians in their AKMs (AK63), so there may have been another motive.

AK47 ASSAULT RIFLE

North Korean (DPRK) Type 58 assault rifle – copy of the Soviet original AK47 Type 3 7.62 x 39mm calibre.

North Korean (DPRK) Type 98-1 provisional illustration. Based on photographic evidence this illustration depicts the 5.45 x 39mm North Korean version of the AK74. The rifle is based on mid-production AK74 and fitted with a right hand side folding stock of North Korean design. Pistol grips appear to be made from plastics and the forearm from possibly laminated wood *(author's collection)*.

In addition the under folding versions are modified by the removal of metal from the metal struts on the under folding stock. This gives the appearance of oval cut-outs along the length of the stock struts and gives the benefit of further reducing weight from the rifle. Magazines have also been modified with 20-round magazines being widely seen in use on Type 68s instead of the usual 30-round type. These magazines are of steel construction.

Type 98
Copy of the AK74 retaining the original 5.45 x 39mm calibre. North Korea is one of the few countries to adopt and retain the calibre outside the counties of the former Soviet Union. The weapon has again been modified by retaining the smooth sided fore end of the Type 68 but most notably the development and adoption of a unique side folding butt. This version is termed the Type 98-1; the metal butt folds to the right of the weapon as opposed to the left hand side on the original Soviet AKS-74. North Koreans produce and use a metal 30-round magazine for the rifle instead of the plastic design produced for the original Soviet version and the plastic designs of the current Russian AK100 series.

The folding stock folds neatly below the charging handle and exposed bolt carrier and appears robust and firm. It has been observed that another unique metal stock design has been produced by the North Koreans – this solution features a stock similar in concept to the under folders of the Type 68, however over folding on top of the receiver cover instead of under the body of the weapon. Photographic evidence has shown this version in service along the border with South Korea but not been seen in North Korean military parades.

Arguably the North Korean-produced AK74 in its Type 98-1 form is a highly credible weapon, right hand folding stock, metal magazine and smooth easy to hold fore end make this version one of the best AK74s ever produced.

Egypt

Egypt has produced the AK for a number of years servicing both the needs of the Egyptian Army and the export market. Egyptian AKs are a straight copy of the Soviet-era AKM with the exception of markings in Arabic.

Misr AKM
Fixed stock copy of AKM.

Misr AKMS
Side folder using a unique single strut bar with triangular stock plate.

German Democratic Republic

The German Democratic Republic known more commonly as East Germany had a long association with the AK. A loyal member of the Warsaw Pact and committed socialist state, the East Germany military followed the Soviets lead on weapon design and procurement. However the AKs produced in East Germany were different in certain design features and they produced a highly credible range of AKs based on the Soviet originals. Making small but useful improvements over the years without changing the basic design function of the weapon, the East German contribution to the development and international success of the weapon was considerable. East German weapons were widely exported to the Middle East, Africa and South America and continue in service around the world today.

East German MPiK selector lever detail.

MPiK
Copy of Soviet AK47 Type 3, but lacks the under barrel cleaning rod. The East Germans produced their own 'pull-through' cleaning kit and therefore had no use for the rod. A small number MPiK rifles were fitted with telescopic sights for use by border guards using a simple side rail mounting on the left hand side of the receiver.

MPiKS
Under-folding stock version of the MPiK.

MPiKM
East German version of the fixed stock AKM. Early versions had wooden stocks and wooden lower fore-arms – the covering on the gas tube was made from plastic. Later versions adopted a unique plastic stock in a brown colouring featuring a 'bubble-effect' pattern. This hollow plastic stock helped to further reduce weight while retaining strength – highly effective. Later models also introduced the AKM type muzzle compensator that was lacking on the early production models.

WORLDWIDE

East German MPiK 7.62 x 39mm assault rifle. A near direct copy of the AK47 Type 3 less the cleaning rod under the barrel. The East Germans issued a separate cleaning kit containing a conventional 'pull though' and so had no requirement for the rod.

MPiKMS-72

Side folding stock version of the fixed stocked MPiKM. The unique East German stock folds to the right of the receiver and is remarkably strong and rigid in position. Early MPiKMS-72 models featured wooden forearms and later models were produced in plastics removing the AKM-type handgrips and replacing them with a flat sided non-slip pattern effect. The mechanism for opening and closing the side folding stock features a lever to operate it, unlike the similar Romanian side folding stock that uses a push-in button system.

MPiAK-74

East German version of the AK74 assault rifle with plastic stock, pistol grip and forearm similar to the later production MPiKM rifles. Magazines were also produced in plastics following the Soviet model in an orange colour. These rifles are of excellent quality and additional accessories included blank firing attachments for the barrel (replacing the standard muzzle compensator) magazine filling tool and clip on 'night sight' attachment for the front sight post. This device provides luminous dots to aid shooting in low level light conditions.

East German MPiKM 7.62 x 39mm assault rifle late production model. The basic layout is a direct copy of the Soviet AKM with changes made to the furniture of the weapon. Stock, pistol grip and forearm are all made from synthetic plastic materials and the stock is hollow. Note the smooth side to the forearm with no finger/hand grips as positioned on the AKM and early production MPiKM rifles. The result of this work eliminated the problem of wear and tear on laminated wooden parts as seen on the original AKM and produced one of the most robust AK designs. Combined with its 7.62mm calibre the MPiKM can be considered one of the best battle rifles ever made.

East German MPiKM 7.62 x 39mm assault rifle early/intermediate model. The first MPiKM rifles produced in East Germany had wooden furniture which over time was replaced by plastics. Note the AKM style forearm complete with finger/hand grips later deleted on plastic forearms. In addition no compensator is fitted which became standard on later production MPiKMs.

East German MPiKMS-72 7.62 x 39mm assault rifle folding stock detail. Note the lever at the base of the mechanism which is moved in order to release the stock to fold.

East German MPiKM 7.62 x 39mm assault rifle. Note plastic stock, pistol grip and fore-end gas tube cover. The plastic hollow stock reduces weight of the weapon and the overall build quality is excellent *(author's collection)*.

MPiAK-74n

MPiAK-74 fitted with a side mounted rail on the left hand side of the receiver for mounting optic sights. The East German Army mounted both night sights and telescopic sights for enhanced accuracy with aimed single shots. The ZFK optic scope could be fitted to the side rail for this purpose and turn the rifle into an effective short to mid-range marksman weapon supplementing conventional sniper rifles.

MPiAKS-74

Side folding stock version of the MPiAK-74 similar in configuration to the earlier MPiKMS-72.

MPiAKS-74n

MPiAKS-74 fitted with side mounted sight rail.

East German MPiKMS-72 7.62 x 39mm assault rifle. This rifle along side the fixed stocked MPiKM represents the high point of AK design – excellent calibre and robust East German construction with plastic furniture – makes this weapon one of the best ever made.

East German MPiAKS-74n compensator and front sight post detail.

East German MPiAKS-74 5.45 x 39mm assault rifle with side folding metal stock folded here to the right of the receiver.

MPiKMS-72 folding stock detail.

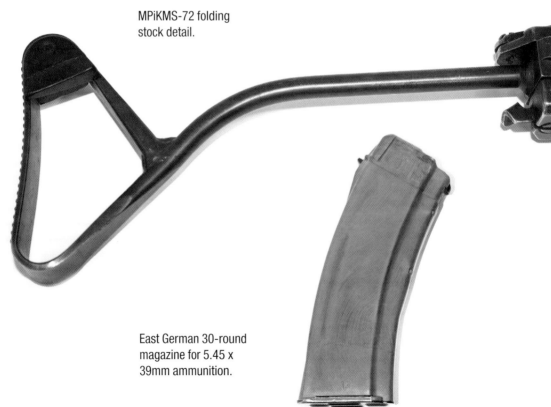

East German 30-round magazine for 5.45 x 39mm ammunition.

East German MPiAKS-74n 5.45 x 39mm assau Featuring a side folding stock, plastic magazine fitted with a sight rail mount on the left hand sic of the weapon. This model is also fitted with a cl on night sight with a luminous dot to aid shootin(low light conditions. All furniture is plastic and lik(the earlier 7.62mm MPiKM the forearm is smooth construction. The folding stock fitted to these rifles extremely tough and provides a firm fire position.

Hungary

Hungary has produced both the AK47 and AKM variants, but not a versio of the AK74 rifle. First production centred around a direct copy of the AK4, but moved to the AKM and distinctive Hungarian versions of the weapon which displayed unique small improvements to the original weapon.

AK-55
Copy of the original AK47 Type 3 in 7.62 x 39mm calibre.

AKM-63
Hungarian development of the Soviet AKM. This model matches the internal and external layout of the original, but the Hungarians made several unique changes which placed this variant ahead of its time in design and production. They replaced the wooden stock and pistol grip with synthetic ones in a light grey/blue colour and completely removed the forearm and upper forearm over the gas tube. They replaced the forearm with a metal supporting structure for a front vertical handgrip, again in a synthetic material. The gas tube was left exposed with no covering.

These changes led to a reduction in overall weight and gave the benefit of a secure two handed firing position. Today nearly all conventional layout

AK47 ASSAULT RIFLE

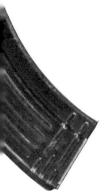

Hungarian AK-55, a copy of the Soviet AK47 Type 3 and the first type of AK produced in Hungary.

Hungarian AKM known as the AK-63D. This weapon retains the four gas vent holes of the AK47 design as found on the Hungarian AK-55. All other features are those of the Soviet AKM with the exception of the wooden furniture. Note the smooth flat sided finish to the side of the forearm as opposed to the raised fingers/hand grips found on the original AKM model. This smooth finish is again a feature retained from the AK-55. Of particular interest is the design of the pistol grip. The size of this grip is somewhat bigger than that found on a standard Soviet plastic AKM pistol grip and provides a greater degree of comfort for the user. The production quality of these weapons is excellent and the type is still in front line service with the Hungarian Army as of 2009.

and some bullpup layout rifles feature the ability to be fitted a forward vertical handgrip – the Hungarians were decades ahead in their thinking and the exposed barrel and gas tube not only reduced weight but aided cooling of the weapon.

The use of synthetic rather than wooden furniture was again well ahead of its time. Only the M16 in the early 1960s was made of a plastic; all others used wood and steel. Today nearly all assault rifles feature some form of plastic/synthetic material in their stocks, lower receivers or pistol grips. Sadly no longer in production, the Hungarians then produced the same AKM in two variants for the Hungarian Army.

AK-63D
From the late 1970s production moved from the AKM-63 to the AK-63D and E models which are largely based on the AKM layout with a flat smooth-sided forearm replacing the vertical handgrip of the earlier model. It lacks the Soviet original grip rails and has four gas vents along the gas tube similar in style to the original AK47 model gas tube. This version is produced with a fixed wooden stock and matching forearm and a straighter, longer pistol grip all in a light-coloured wood. Construction quality is excellent and the rifle was exported outside of Europe from the 1980s onwards.

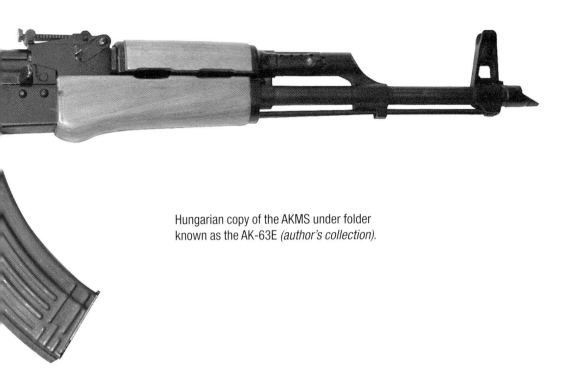

Hungarian copy of the AKMS under folder known as the AK-63E *(author's collection)*.

Close up of receiver, pistol grip and magazine of the AK-63E. With the exception of the pistol grip design this is a copy of the Soviet AKMS original. The selector lever is in the upper safe position – in this position the weapon cannot be fired or the charging handle pulled fully back *(author's collection)*.

Hungarian AK-63E with selector lever moved to the lower position for semi-automatic fire *(author's collection)*.

Hungarian AK-63E – top view of the ribbed receiver cover and rear sights. Note the rear sight is graduated from 100 metres to 1,000 metres as per the Soviet AKM original. Also the rear 'A' position is a battle range sighting *(author's collection)*.

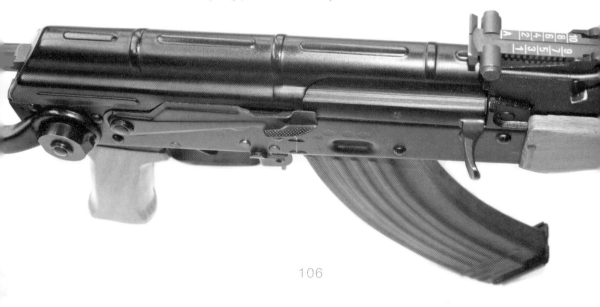

AK-63E

Under folder version of the above model featuring a steel under folding stock similar in design to the original AKMS. Both types are in service today with the Hungarian Army though production in-country has ended.

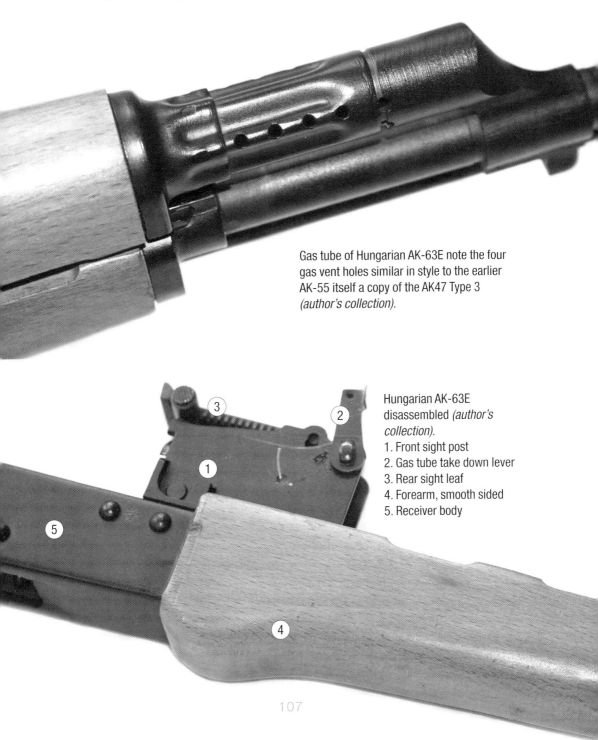

Gas tube of Hungarian AK-63E note the four gas vent holes similar in style to the earlier AK-55 itself a copy of the AK47 Type 3 *(author's collection)*.

Hungarian AK-63E disassembled *(author's collection)*.
1. Front sight post
2. Gas tube take down lever
3. Rear sight leaf
4. Forearm, smooth sided
5. Receiver body

Iran

The Iranian Defence Industries Organisation (DIO) produces several versions of the AK which appear to be a hybrid of the AK47 and AKM (similar to late and current production Chinese Type 56). Iranian-produced AKs have a good reputation for quality and have found export markets in limited numbers.

KLS
Fixed stock.

KLF
Under folding stock.

KLT
Side folding stock.

All versions feature plastic pistol grips, AKM-style receivers, AK47-style gas tubes with four gas vents and AK47-style smooth forearms. Front sights are Chinese-style fully enclosed.

Iraqi M70B1 7.62 x 39mm assault rifle.

Iraq

Prior to the invasion of Iraq little was know about AK production in the country and few people cared. However, the conflict that flared between rival groups and the US and Allied forces found Iraqi AKs at the heart of the war. Iraq produced its own rifles based on the Yugoslav M70B1 with fixed wooden stocks and metal under folders. Apart from Arabic markings, the Iraqi weapon and the Yugoslav original are near identical.

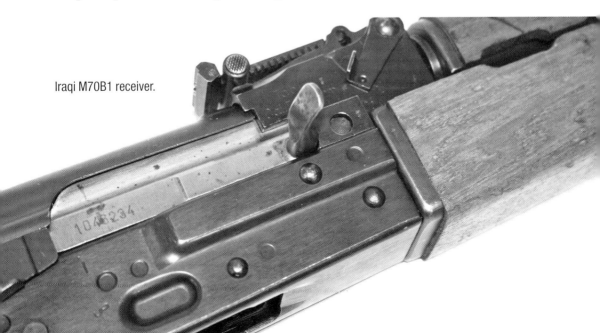

Iraqi M70B1 receiver.

Iraqi M70B1 receiver.

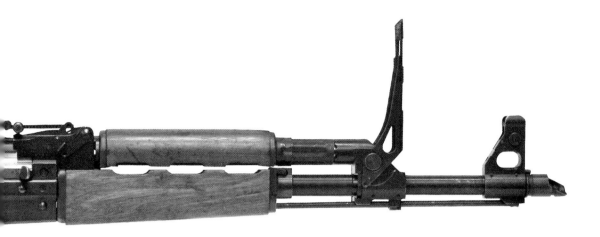

Iraqi M70B1 with front grenade aiming post raised.

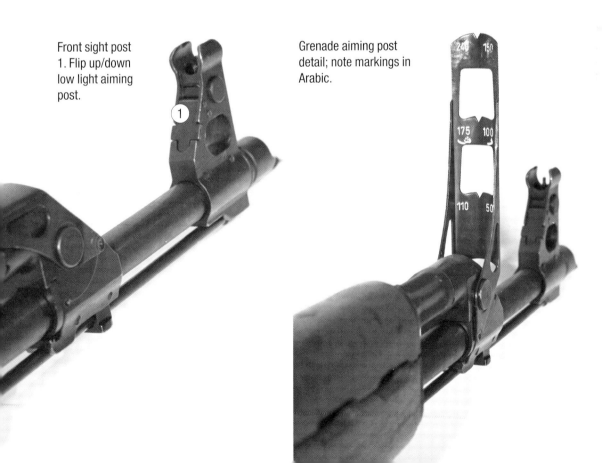

Front sight post
1. Flip up/down low light aiming post.

Grenade aiming post detail; note markings in Arabic.

Pakistan

Pakistan produces a copy of the AKMS known as the PK10. 7.62 x 39mm.

Poland

Poland has produced variants of all three major incarnations of the AK family, mostly produced at the Random factory which to this day produces small arms based on the AK's internal design. Poland has been a massive producer of the weapon and was a major exporter during the communist era. Polish weapons have a sound reputation and the developments made in the Tantal showed a move away from the original Soviet models. Today Poland produces a number of designs based on the AK in standard NATO 5.56 x 45mm and continues to show growing design innovation. The versions listed below are now out of production but can be found all over the world and represent the direct Polish versions of the Soviet originals. Polish AKs continue to serve with the Polish Army today though are being replaced slowly with variants built to fire the NATO 5.56 x 45mm round.

Polish KbK AKMS 7.62 x 39mm calibre, copy of Soviet AKMS. This particular model was built for export to the Arab world and features a rear sight with Arabic markings. Note the plastic pistol grip and finger/hand grips on the forearm. Apart from the Arabic markings this model is the same as the Polish version still in use by the Polish Army.

KbK AK
Copy of the Soviet Type 3 AK47 with wooden furniture – 7.62 x 39mm calibre. Produced 1957–1958.

KbK AKS
Copy of Soviet Type 3 under folder AKS47 – 7.62 x 39mm calibre. Produced 1957–1965.

KbKg
Version of the KbK AK out fitted for the launching of rifle grenades. The rifle features a screw-on spigot barrel extension and a padded stock pad to absorb the kick of the grenade when firing. Both anti-tank and high explosive grenades were developed for the weapon and Poland along with Hungary and the former Yugoslavia developed grenade launching solutions for the weapon in the 1960s.

KbK AKM
Direct copies of the Soviet AKM complete with wooden furniture and muzzle compensator – 7.62 x 39mm calibre. Produced 1966–1972.

KbK AKMS
Direct copy of the Soviet AKMS under folder – 7.62 x 39mm calibre. Produced 1972–2000.

KbK wz 1980
A development of the under folding KbK AKMS, this rifle is fitted with a fire selector on the left hand side of the weapon in addition to the standard AK selector lever on the right. The design was possibly made to meet export demands from the Middle East – 7.62 x 39mm calibre.

KbK wz 1988
Known as the Tantal, this version was produced in the Soviet 5.45 x 39mm calibre and was based on the KbK AKM variant. The design is not a copy or a development of the Soviet AK74 though it shares the same calibre. All examples are fitted with a side folding metal stock similar to the East German MPI AKMS-72 model. In addition the weapon is fitted with a unique polish muzzle compensator different in configuration and use to the AK74.

PM Md 63 7.62 x 39mm assault rifle with fixed wooden stock and vertical front hand grip.

Romania

The Romanians began producing AKs from the late 1960s and continues to do so today, incorporating some unique features. They are most noted for mainly having an additional vertical handgrip built into the lower forearm to improve control when firing. Like the Hungarians who developed a similar solution with their AKM-63 model, this style of vertical second hand/pistol grip is now almost standard in current small arms design. Romanian AKs have been and are still widely sold around the world.

PM Md 63
Copy of the AKM, but incorporating a curved forward handgrip built into the lower forearm. Laminated wooden stock and forearm furniture. 7.62 x 39mm calibre.

PM Md 65
Under folding stock version of the PM Md 63.

PM Md 65 7.62 x 39mm assault rifle with folding metal stock.

PM Md 86 5.45 x 39mm assault rifle fitted with folding metal stock. Romanian AKs in 5.45 x 39mm calibre are of a unique design incorporating features from the Soviet AKM and AK74.
1. 45 degree gas block
2. Up-turned charging handle
3. Compensator – thinner in construction to Soviet/Russian AK74s
4. Vertical front hand grip
5. Right hand side folding stock
6. Steel magazine

PM Md 86

Romania's version of the AK74. Heavily influenced by the AKM, this rifle in 5.45 x 39mm calibre has a number of distinctive features.

- Laminated wooded lower forearm incorporating forward handgrip.
- 45 degree angle gas block (like AKM).
- Bakelite pistol grip.
- Upturned charging handle – changing handles turned a few degrees upwards to aid use when stock folded and allow folding stock to fall into folded position unhindered.
- Right hand folding stock - similar in pattern to East German MPiKMS-72. Button open/close system with different push.
- 30-round steel magazines.
- 3-round burst option on selector lever.
- Compensator on muzzle – different pattern to Soviet AK74 original – thinner profile.

Also produced without forward handgrip, this model incorporated a lower forearm design similar to East German MPiAK-74.

PM Md 86 with 30-round magazine removed.

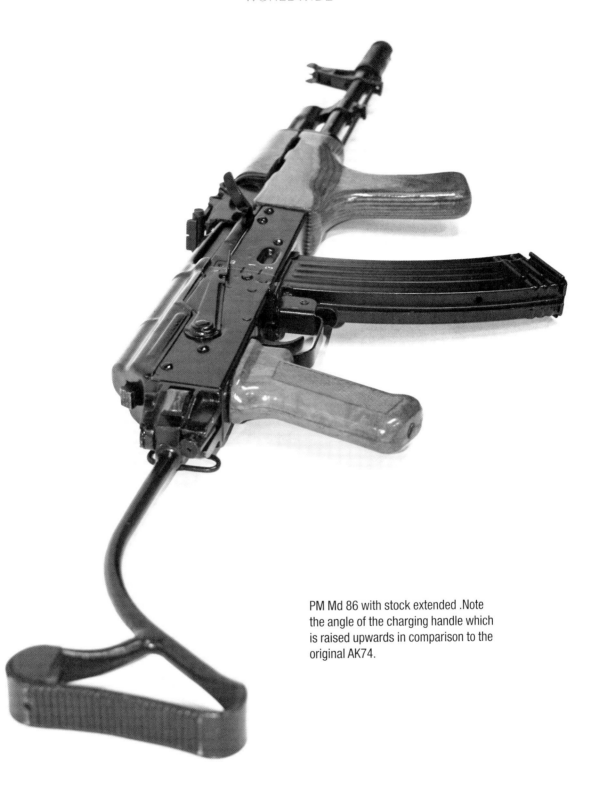

PM Md 86 with stock extended .Note the angle of the charging handle which is raised upwards in comparison to the original AK74.

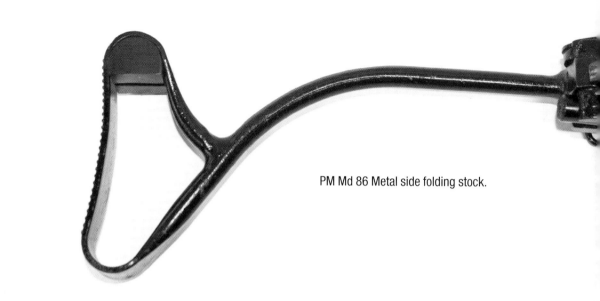

PM Md 86 Metal side folding stock.

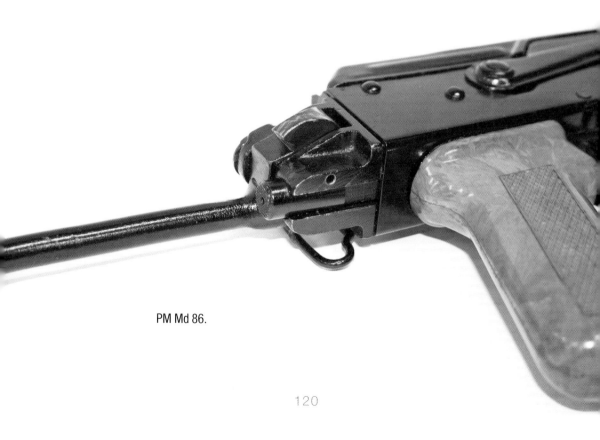

PM Md 86.

WORLDWIDE

PM Md 86 charging handle detail.

PM Md 86 muzzle compensator.

PM Md 90
Based on the AKM this model features a right hand side folding stock similar to the version fitted to the PM Md 86. 7.62 x 39mm calibre.

PM Md 86 Receiver detail.

PM Md 90 7.62 x 39mm assault rifle.

30-round magazine for PM Md 86 assault rifle.

Russia

Production of the AK continues today in its 'home' of Izhevsk. Indeed, Russian weapons are once again becoming popular, with export sales success being found with the latest incarnation – the excellent AK103.

AK103
The AK103 is now the 'high point' of the AK – it combines the best of the AKM and the AK74. Retaining the 7.62 x 39mm round and the muzzle compensator of the AK74, this weapon represents all of the best design innovation of Kalashnikov team's work. It is likely that this AK will continue to be sold around the globe for many years to come.

AK74M
Current production AK74 with black furniture in 5.45 x 39mm calibre.

WORLDWIDE

M70B1 assault rifle 7.62 x 39mm calibre and featuring a mix of AK47, AKM and unique Yugoslavian features such as the built-in rifle grenade sighting post over the gas tube area and pistol grip design.

Serbia (former Yugoslavia)

The now defunct Yugoslavia began work on AK47 variants from the early 1960s developing the M64 series and M70 rifles on milled receivers. This work cumulated into the highly regarded M70B1 and M70AB2 assault rifles which are still offered for export today by manufacturers in Serbia.

M70B1
This rifle is a hybrid design taking features from the AK47, AKM and Kalashnikov's light machine gun version the RPK. Retaining the 7.62 x 39mm calibre the weapon features are:

- Teak wood stock and forearm.
- Plastic pistol grip unique to Yugoslav rifles.
- Larger trunnion – a feature taken from the RPK light machine gun.
- Stamped metal receiver 1.5mm thick – the AKM original is 1mm thick.
- Non-chromed barrel – unusual for an AK and a potential rust issue if not cleaned properly.
- Stock has a thick rubber pad attached in order to absorb recoil shock.
- Flip up luminous dot night sights on both the front and rear sights.

- Built-in grenade sight which shuts off gas flow via a valve between the barrel and gas chamber to enable the launching of rifle grenades.
- Small push-in button at rear of receiver/receiver cover – enables the recoil mechanism to be pushed forward, allowing removal of the receiver cover.
- Last round bolt hold opening device by means of a bespoke magazine design – the follower within the magazine raises up into the pathway of the bolt/bolt carrier and stops its forward movement when the last round is fired.

The larger trunnion and thicker receiver all add to the strength and durability of the design but weight is affected; the rifle weighs in at 4.2kg (9.3lbs). The lack of chroming on the barrel could lead to rusting if not cleaned regularly and the push-in button for the removal of the receiver cover is a piece of unnecessary over-engineering. The magazine incorporating 'hold-open' design is useful, though when the empty magazine is removed from the weapon the bolt/bolt carrier moves forward under the pressure of the recoil mechanism (spring). Many people regard the M70B1 as one of the best variants produced, but in the author's opinion its extra weight, lack of barrel chroming and unnecessary features limit its worth.

Sudan

Sudan produces a copy of the Chinese Type 56 AK47 known as the MAZ. It is of the wooded fixed stock, under folding spike bayonet type. At the time of writing no Sudanese under folding version has been found by the author. It is likely that the Iranians are instrumental in aiding Sudanese production.

Close up of M70B1 forward receiver, bo carrier and charging handle. Of note is construction of this forward area of the receiver featuring the enhanced trunnic

AK47 rear sight detail. Note sight marked to 800 metres. AKMs are sighted to 1,000 metres.

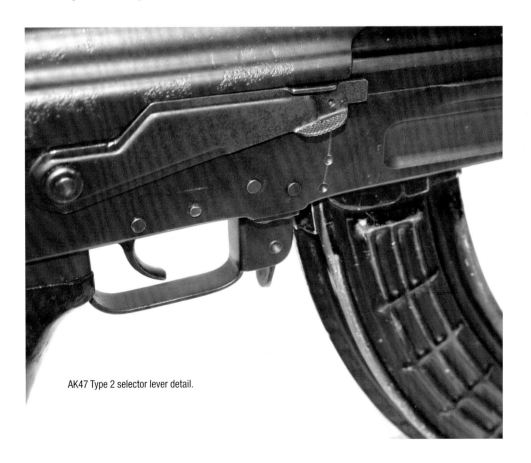

AK47 Type 2 selector lever detail.

Soviet AKM charging handle detail.

AKM gas tube.

AKM underside disassembled.

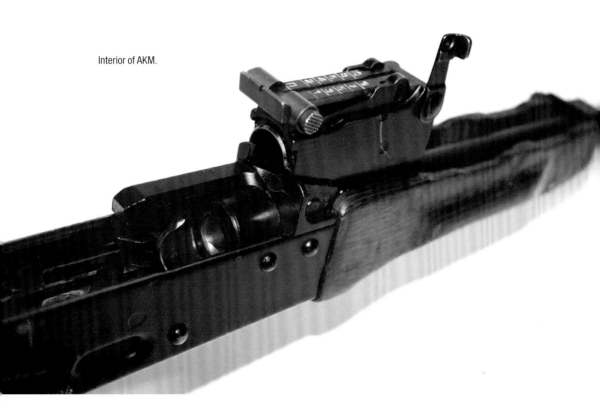

Interior of AKM.

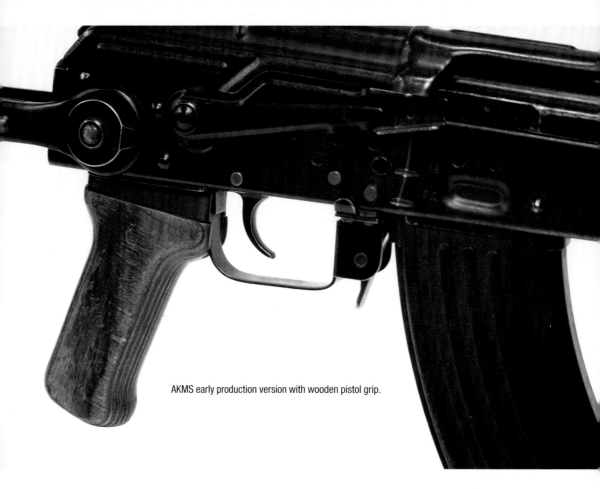

AKMS early production version with wooden pistol grip.

Rear view of AKMS.

AK74 muzzle compensator.

AK74 with skeleton folding metal stock extended into firing position. The button to release the stock from this position for folding is positioned to the rear of the receiver in front of the stock mechanism.

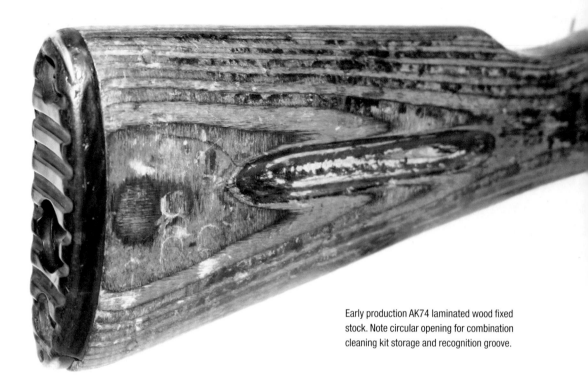

Early production AK74 laminated wood fixed stock. Note circular opening for combination cleaning kit storage and recognition groove.

AKM combination tool/cleaning kit.
1. Screwdriver blade
2. Bore brush
3. Drift
4. Jag
5. Combination tool cap which also acts as a bore guide when combined with the cleaning rode
6. Combination tool which doubles as a handle to the cleaning rod
7. Spare pin for use within the firing mechanism

Close up of Type 56-2 unique pistol grip and side folding stock mechanism.

East German MPIAKS-74n with siding folding stock folded to right hand side of the receiver.

Albanian Type A assault rifle – a near copy of the Chinese Type 56. Note absence of any form of dimple or cut-out on the side of the receiver above the magazine. Other features include smooth top to receiver cover (not ribbed like AKM), non parkerised bolt carrier, wooden furniture and fixed spike bayonet.

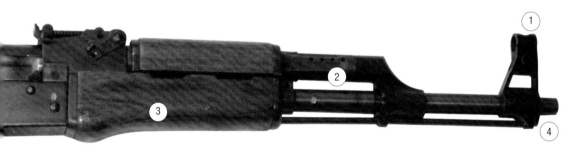

Type 56-1 assault rifle *(author's collection)*.
1. Enclosed front sight
2. Gas tube featuring four gas vent holes
3. Forearm
4. Unique muzzle break
5. 30-round magazine for 7.62 x 39mm ammunition. Chinese magazines have no spine running down the curved back of the magazine's body and is a recognition feature of Chinese produced AK type magazines.
6. Metal under folding stock
7. AKM style dimple
8. Smooth receiver cover

Hungarian AK63F with selector lever moved to the middle position for full automatic fire *(author's collection)*.

Hungarian AK63F with receiver cover removed *(author's collection)*.

1. Recoil mechanism
2. Bolt carrier.
3. Wooden pistol grip of Hungarian design, larger in size than Soviet AKM pattern
4. Selector lever
5. Recoil mechanism lug
6. Magazine release catch
7. Magazine well
8. Charging handle
9. Rear sight
10. Receiver cover, ribbed reinforced

Front sight post detail with low light aiming post raised to show luminous dot aiming mark.

East German 30-round magazine for 5.45 x 39mm ammunition.

PM Md 86 with stock folded to the right hand side of the weapon.

M76 sniper rifle with magazine removed and showing the sight mounting bracket on the left hand side of the weapon.

Mikhail Kalashnikov holding the weapon that made him famous. (*Novosti*)

Above: East German soldiers cleaning their weapons. While still reliable if not cleaned, the AK's longevity is dramatically improved with regular maintenance. *(US Army)*

Left: A soldier from the Chinese navy with a Type 56. *(J.P. Husson)*

Troops on patrol armed with AK47s and an RPD light machine gun.

A US Army Ranger on patrol in Vietnam armed with an AK, which offered greater reliability than the US M16. *(US Department of Defense)*

Mujahideen guerrillas with AKM rifles during the Soviet-Afghan War. The Mujahideen acquired their weapons through CIA-sponsored supplies or by capturing them from the Soviets. *(TASS)*

A Kuwaiti soldier during the Gulf War with an unloaded AK. *(US Department of Defense)*

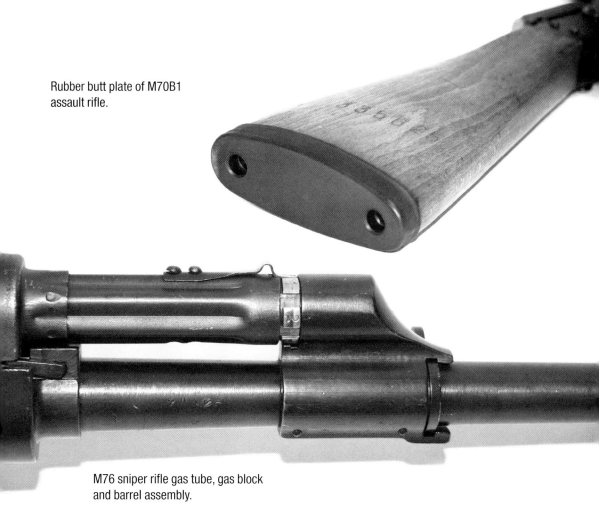

Rubber butt plate of M70B1 assault rifle.

M76 sniper rifle gas tube, gas block and barrel assembly.

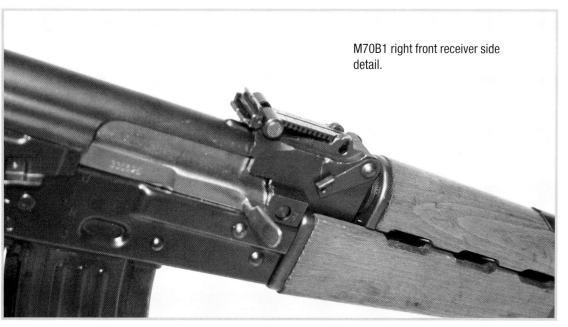

M70B1 right front receiver side detail.

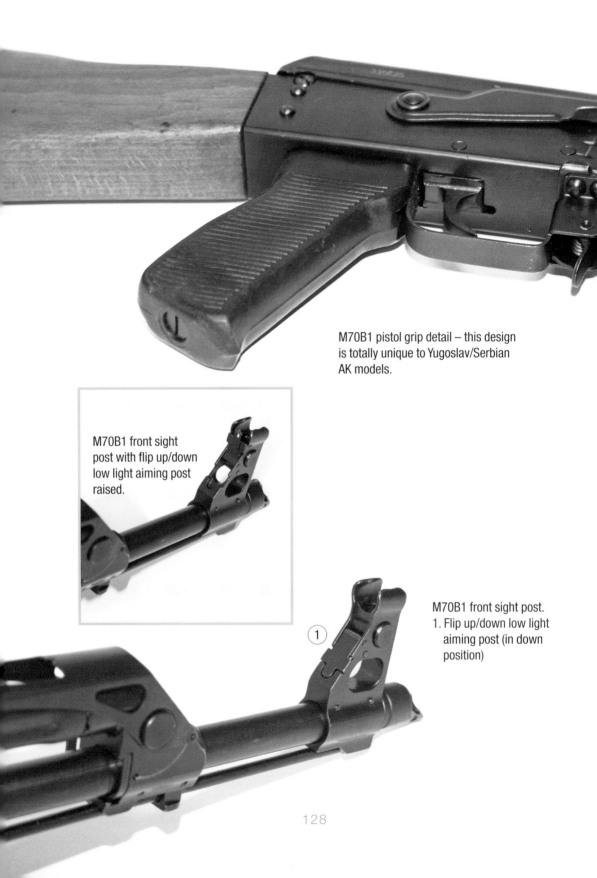

M70B1 pistol grip detail – this design is totally unique to Yugoslav/Serbian AK models.

M70B1 front sight post with flip up/down low light aiming post raised.

M70B1 front sight post.
1. Flip up/down low light aiming post (in down position)

M70B1 underside view of receiver and magazine. Yugoslav/Serbian magazines for the M70B1 have a bolt 'hold open' feature whereby when the last round leaves the magazine the follower protrudes at a height which blocks the bolt and bolt carrier from moving forward. Once the empty magazine is removed from the weapon these working parts will push forward under the pressure from the recoil mechanism.

Vietnam

Details of Vietnamese production are unknown. What is known is that the Vietnamese have been very skilful in rebuilding AK47s, AKMs, Type 56s and other variants in order to keep them in service. Versions of the weapon have been seen containing parts from various different sources rebuilt into one working serviceable gun. This is a testament to the weapon's design and the skills of the Vietnamese at recycling weapons in order to maintain stocks without the expenses of new small arms purchases.

Worldwide Design Influence

The AK's design influence on assault rifles produced after the Second World War is enormous. Many nations took the basic design and modified it for their national needs leading to a massive range of rifles all bearing the internal and some times external characteristics of the weapon. The Israelis, Finns and South Africans all used the AK's layout both internally and to a large extent externally on their rifles; the Finns produced the finest example with the RK95 service rifle in use today with the Finnish Army. The Finns also retained the 7.62 x 39mm calibre whereas all others went for the US 5.56mm option. Finland's RK95 is exceptional – very well made and solid and largely an AK in all but name.

East Germany built upon the success of the AK/MPiKM design by producing a version aimed at the export market outside of the Warsaw Pact allied countries. Basing the weapon on the western 5.56mm round, they created the STG 942, often known as part of the Wieger STG 940 series. The STG 942 featured a folding stock, and was exported to Peru with limited success. The internal structure remained pure AK, but the magazine, barrel, gas tube, front aiming post and fore end furniture were heavily modified. With the fall of the Berlin Wall production ceased; however in the recent years the design has been resurrected by the Romanians.

During the 1980s South Africa was embroiled in a series of border conflicts with Namibia and Angola. Looking for a replacement for the long-serving R1 rifle (a local copy of the Belgian FAL rifle) the South Africans looked to the Israelis and their work on the AK design. The Israeli Galil rifle was based on the AK but chambered to the 5.56mm round. Taking this design the South Africans modified it and produced the R4.

Outwardly similar to the Galil ARM the R4 features a longer polymer stock, a polymer 35-round magazine and modified fore end furniture. It retains the Galil underfolding bipod under the barrel, separate gas tube, self-luminous tritium dot sights for use in low light conditions and a right hand side folding stock. Interestingly the charging handle is turned up into a vertical position to allow use with the left hand. Selector levels are also provided on both sides of the weapon with a standard AK-style sector on the right hand side and smaller selector switch on the left above the pistol grip. Solid and reliable, the R4 is a classic battle weapon; its major limitation is that it is designed to use the older M193 5.56mm round rather than the more effective current SS109 5.56mm round.

Other examples of the influence can be found is less obvious areas as is the case of the Singaporean SAR21 assault rifle. This design is in the 'bullpup' configuration with the magazine positioned behind the pistol grip reducing overall length. Upon looking inside the SAR21 however, we find a bolt carrier and gas piston rod all made in one complete piece similar in layout to the AK design. In addition the concept behind the rifle matches many of the AK's features – it's very simple to disassemble and reassemble, has few parts and though fitted with an optic sight this sight is built directly onto the upper receiver and barrel. This set-up means that

SAR21 5.56mm assault rifle. Manufactured and designed in Singapore in a bullpup configuration and featuring a built in x1.5 optic sight. The weapon also has a laser red-dot projector visible in the front of the fore-end hand guard. The SAR21 is easy to use, reliable and accurate.

SAR21 assault rifle field stripped for cleaning. The weapon can be disassembled and re-assembled in seconds and without tools. Note that the gas piston is directly connected to the bolt carrier in a similar manner to an AK.

the rifle can be pre-zeroed at factory level and therefore requires little to no further zeroing by the shooter. This feature reduces training time and gives greater confidence to the user. Given that the Singapore Army is mainly made up of young conscripts, the ease of use combined with the built-in sight makes for a rifle ideal for a 'people's' army.

Many assault rifle designs today feature multiple rail systems which allow the user to switch different sights and other targeting devices. The SAR21 has no rail system and is kept simple with a sight that stays on the rifle and can't be removed by the user. The advantage of this – especially when the soldier is likely to be a conscript – is that parts cannot be lost on the battlefield and the zero of the rifle is not affected by a soldier removing sights unnecessarily. Israel followed a similar concept with the development of the Tavor assault rifle, which also has the sight built directly on the weapon. The SAR21 and the Tavor were designed along the same principles as the AK, although they are let down by their calibre. The Chinese have been more successful with their QBZ 95 rifle. This weapon is also a bullpup with internal working parts largely influenced by the AK with bags of space between the recoil mechanism and the bottom of the receiver. The advantage of the Chinese solution however is its larger calibre of 5.8mm. Kalashnikov's work will continue to influence assault rifle design for a long time to come because there is currently nothing better.

East German STG 942 with the 30-round 5.56mm ammunition magazine removed.

STG 942 built for the export market, a number of which were sold to Peru.

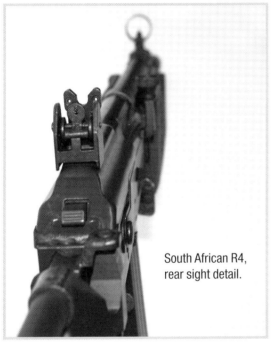

South African R4, rear sight detail.

STG 942 front leaf sight detail.

AK47 ASSAULT RIFLE

South African R4 with stock folded to the right hand side of the receiver. Note the AK-style selector lever.

R4 front sight detail.

CHAPTER SIX

SNIPER VARIATIONS

The original Soviet Army requirements for a new assault rifle in the late 1940s did not ask for the development of a sniper variant and the resulting Kalashnikov design did not incorporate a version suitable for battlefield sniping. All too often the western press link the AK47 to the Soviet SVD sniper rifle, but in fact the two designs came out of two different Soviet Army needs and different design bureaus.

The Soviet Army (then Red Army) fought during the Second World War with the excellent M1891/30 sniper rifle fitted with PE or PU scopes. The rifle served the Army well and the Soviets built up a cult around the lone sniper in defensive urban battles such as Stalingrad. Following the war and the introduction of the AK47 and then the AKM, the army issued a new requirement for an improved sniper rifle based on the experience of wartime sniping and the needs of modern mechanised warfare. This led to an open competition similar to the one that produced the AK47 and resulted in the development and adoption of the semi-automatic SVD sniper rifle.

The SVD rifle is not and was never an AK. However, a number of countries that adopted the AK47 design produced sniper rifles based fully on the AK working parts. The only internal and external changes were the barrel length and semi-automatic firing mode. The rifles produced were excellent short to medium range battlefield sniping weapons with the former Yugoslavian and Romanian versions still in manufacture today and widely used.

M76 Sniper Rifle

Designed and produced in the former Yugoslavia and still produced in Serbia today. It has also been reported that a copy version has been produced in the DPRK though this is unverified by photographic evidence.

The internal working parts of the M76 are pure AK simply modified to reflect the longer barrel and corresponding gas tube. Interestingly the rifle is cambered in its original configuration for a 7.92 x 57mm round – a larger calibre then most of its contemporaries. Unlike AK-type assault rifles the weapon features semi-automatic fire mode only.

As standard the M76 is fitted with a x4 magnification scope, also known as the M76 scope, and is attached to the left hand side of the rifle via a slide-on rail system. Larger calibres for infantry sniper rifles are currently all the rage, giving both extended range and maximum killing power against hardened targets (troops behind light cover and troops wearing body armour for example).

The M76 is an excellent example of a sniper weapon of this type – designed to be a rifle deployed within the lowest level of infantry organisations – section, platoon and company level. Today the 'designated marksman' is commonplace within western armies – a soldier deployed at section or platoon level and armed with a large calibre sniper weapon to counter long range targets. The M76 fills this role well and still retains a degree of lightness and ease.

SNIPER VARIATIONS

M76 sniper rifle.

M76 sniper rifle with magazine removed.

AK47 ASSAULT RIFLE

Calibre	7.92 x 57mm
Muzzle velocity	730 metres per second
Weight	4.6kg
Effective range	Up to 800 metres
Magazine capacity	10 rounds

M76 sniper rifle.

SNIPER VARIATIONS

Romanian PSL sniper rifle 7.62 x 54 mm. This sniper rifle combines the true 'AK' internal working parts set-up with the sniper rifle features of scope, barrel length and furniture to aid firing position hold. Its outward appearance is very similar to the Soviet/Russian SVD sniper rifle and shares the same calibre. However the internal mechanism of the SVD is very different to that of the AK47 and AKM and works on a short stroke gas pistol system. The PSL on the other hand is very much an AK on the inside and disassembly and assembly of the rifle is very similar to the AKM with the addition of removing the optic sight from its left hand side slide mounting.

PSL Sniper Rifle

Designed and still produced in Romania, externally the rifle resembles the Soviet SVD sniper rifle, but internally it is based on the AK. Chambered for the 7.62 x 54mm round as used within the SVD and equipped with a x4 magnification scope known as the LPS Tip2. The scope is fitted to a slide-on rail system positioned on the left hand side. This rail also allows the fitment of any sighting systems such as night sights equipped with a compatible rail mount.

Calibre	7.62 x 54r mm
Weight	4.3kg
Effective range	Up to 800 metres
Magazine capacity	10 rounds

Tabuk Sniper Rifle

Designed and produced in Iraq during the Saddam regime, with aid from the former Yugoslavia in the form of production machinery. The weapon is based on the in-country Tabuk assault rifle which is a copy of the

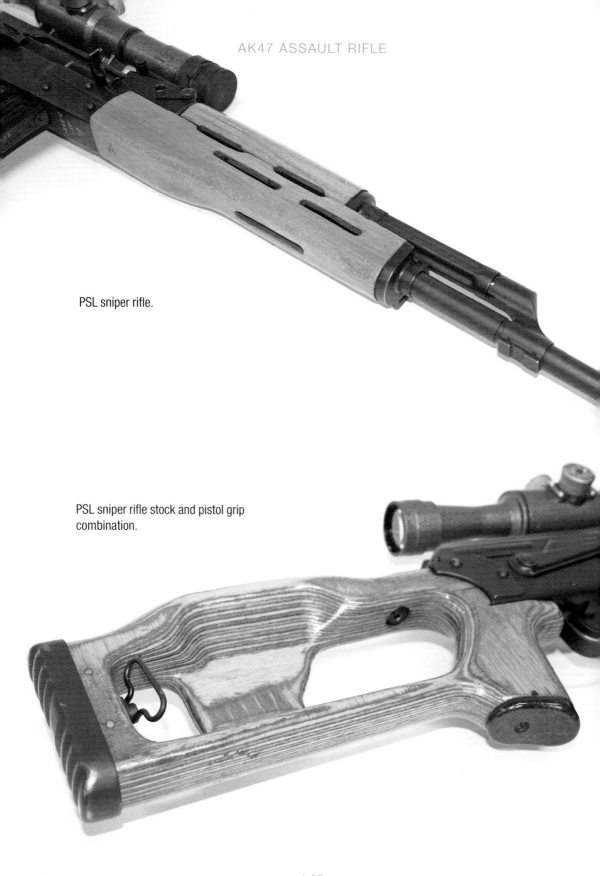

PSL sniper rifle.

PSL sniper rifle stock and pistol grip combination.

SNIPER VARIATIONS

Yugoslavian M70B1 model. It features an extended barrel, skeleton-type wooden stock, selector lever and two positions only – safe and semi-automatic. The rifle retains the standard 7.62 x 39mm calibre rather then a larger cartridge with a calibre suitable for extended ranges. The weapon is effective in its role up to 400 metres by virtue of its extended barrel length and can be fitted with a variety of sights on its side rail sight mount.

PSL sniper rifle muzzle break, designed to remove/hide flash.

PSL sniper rifle.

PSL sniper rifle.

CHAPTER SEVEN

AMMUNITION AND MAGAZINES

The 30-round magazines designed for the AK47 and AKM are without doubt the finest in the world. Like the weapon itself, their simple yet inspired designs have been key to the AK's success. Poor magazine design, manufacture and maintenance is one of the main causes of stoppages in military assault rifles. The M16 in its first incarnations is a great example of getting it all wrong. It was first issued with a 20-round straight up and down magazine, but the spring inside was not strong enough for sustained use and would fail to feed rounds. The soldier's solution was to load the magazine with only sixteen rounds to compensate for its weakness. In addition the US armed forces only issued 30–round magazines in any numbers after the Vietnam War when it was too late for the poor US infantrymen.

The AK magazine is a totally different story. Its now familiar curved shape follows the line of 30 rounds stacked 'nose down' and unlike western designs based on the M16 30-round magazine is made in one continuous curved line with no 'kink' that could cause a stoppage. The reason why western designs are like they are is simple – they were born out of the fact that the M16 was originally designed for a 20-round straight box magazine and the magazine well on the lower receiver is designed with this in mind. The AK47 was designed from the start to have a 30-round magazine and a magazine well that does not extend out of the bottom of the weapon, therefore not limiting design options.

Original AK47 magazines made from steel were smooth sided. Often referred too as 'slab sided' these rare early types were replaced by the now

AK magazine disassembled for cleaning *(author's collection)*.

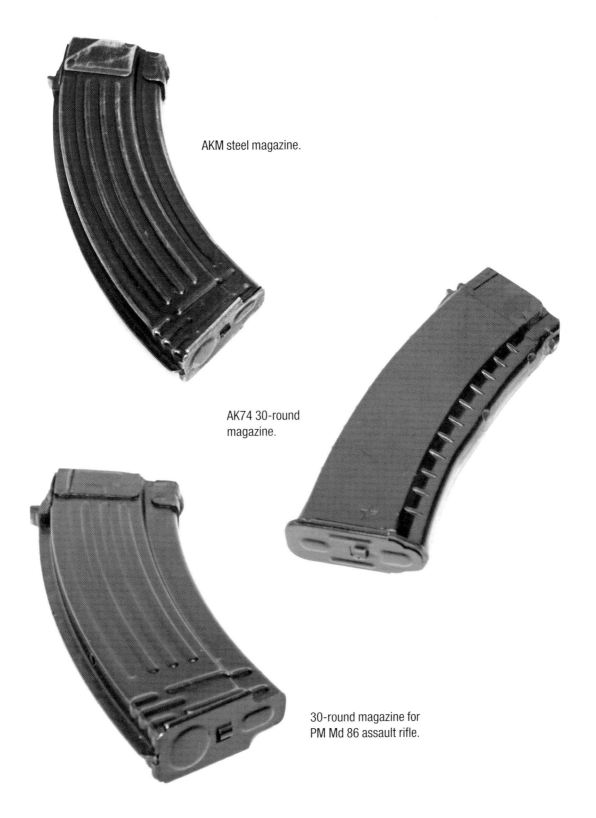

AKM steel magazine.

AK74 30-round magazine.

30-round magazine for PM Md 86 assault rifle.

familiar rib-sided models, again made from steel. With the introduction of the AKM in 1959 the Soviets produced the same type of magazine in an orange coloured plastic (smooth sided) and did the same with the introduction of the 5.45 x 39mm AK74. Though both were successful the preference went to magazines built from steel rather then plastic. Most AK producers today produce steel magazines, though those built in Russia for the excellent AK103 are made out of black plastic.

Several other nations produced plastic magazines for their national variants, notably the East German 5.45 x 39mm magazines for the MPI AK74 and the Bulgarians today produce world class AK magazines in plastic. Bulgarian magazines are partially favoured by users and are clearly identified by the criss-cross pattern on the outside of the magazine referred too as the 'waffle' pattern.

Much has been said by the author in this book regarding the calibre of the AK, namely the 7.62 x 39mm ammunition and the later 5.45 x 39mm round. Lethality is what it is all about – the more energy that can be deposited or 'dumped' into the human body the more lethal the ammunition and therefore the weapon. A high velocity round can fly quicker to the target then lower velocity rounds but use the energy in their speed in medium to long ranges – leaving little energy or power to do damage to the target. In addition at short ranges they can go through a human target and out the other side, and should they not hit a vital organ the target can fight on. Stories from battlefields show this frequently happens, and a soldier may have to fire many rounds into a target before stopping it dead.

The Soviets found the balance between muzzle velocity, calibre and construction composition in both the 7.62mm and 5.45mm rounds. Experts around the world still argue about the merits of the US 5.56mm calibre rounds, but the fact is that the Russians designed the best ammunition for killing people.

M43

The M43 basically consists of a projectile (the bullet) casing, propellant charge and a primer. The casing of the round is normally produced from copper washed or lacquered steel being coloured dark green, brown or grey. Seals at the join between the casing and the bullet are red in colour and the purpose of this seal is to aid a tight join between the two parts.

The M43 round was and still is produced in three basic types:

- Conventional 'ball' round with a steel core – known as the 'PS' round – muzzle velocity for this round is 710 metres per second.
- Tracer round – visible out to 800 metres – identified by its green colour nose.
- Armour piercing incendiary round (API) – known as the 'BZ' round. This round has a copper – zinc nose cap, steel core, lead liner and an incendiary substance contained within the bullet. It is identified by having a black coloured nose above a red stripe. Armour penetration is up to 7mm at 200 metres.

This type of 7.62 x 39mm ammunition has been produced in nearly 50 countries since the late 1940s and continues to be widely manufactured throughout the world today.

M74

Soviet/Russian 5.45 x 39mm ammunition is known as the M74 round or commercially the 5N6 round. This round of ammunition has a number of features which combined with the AK74 rifle make it highly lethal. Muzzle velocity is 900 metres per second, which combined with the rifling effect of the AK74's barrel gives a very flat trajectory of flight to its target. The bullet itself consists of several parts – a steel core followed by a lead inlay followed by a hollow tip. This hollow tip combined with the centre of gravity being at the rear of the bullet causes the round to tumble when it strikes a human target. The effect of the tumble is to 'dump' its energy into the body.

The 5.45 x 39mm round has undergone a number of updates to improve its effectiveness in response to changing battlefield conditions. With personal body armour becoming the norm for armies, ammunition is required to penetrate this new defensive measure. At least three other types of round have been made:

7N10 (known as the 'PP' round)
Improved performance against body armour, has a small hardened steel penetrator. Purple seal. Two models of this round exist. The 1992 model retains a small hollow air cavity similar to the earlier 7N6. The 1994 model has no hollow air cavity.

7N22 (known as the 'BP' round)
Has a high carbon steel core and can penetrate 5mm of armour at 250 metres. Black tip and red seal.

7N24 (known as the 'BS' round)
Improved 7N22 model and capable of penetrating 5mm of armour at 350 metres. Black seal.

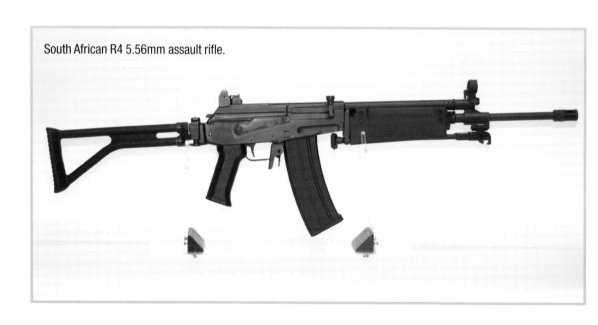

South African R4 5.56mm assault rifle.

SUMMARY

It works, it does the job and is the standard by which other assault rifle designs are judged. The original choice of calibre in 7.62 x 39mm was the right choice and still is today. Many will argue the merits of smaller calibres such as the standard NATO 5.56mm round, but they do not measure up. The Soviet Union's fusion of two design greats – Kalashnikov and his rifle and the 7.62 x 39mm round of ammunition – produced the right result.

However, the AK47 is no stranger to criticism. It has been accused of inaccuracy, poor craftsmanship, a noisy selector lever and of having loose fitting parts such as the upper forearm and gas tube. All of these charges are partially true but miss the point. If the upper forearm gas tube is loose it doesn't matter because the weapon will still work so long as the gas tube itself (contained within the upper forearm) is not blocked. The weapon is accurate enough within battle ranges and comments made about poor workmanship can be dismissed. The weapon is highly reliable and continues to work in conditions where other rifles supposedly built to high standards would stop functioning in half the time.

The quality of the Russian AK103 and Bulgarian AR-M1F are excellent and outshine other weapons on the small arms market today which are often over engineered and chambered for a round that no one believes in. The Bulgarian AR-M1F is solid, well made, has no fancy gimmicks on it that can break and fires a round of ammunition that a soldier can trust – a 7.62 x 39mm round. It is hard to see an end to this weapon and wherever men want to make war someone will make an AK. The lessons of history are clear – keep it simple, make it tough and use a big bullet! The AK47 is the design classic of the gun world – it is designed to kill and that's what it does.

South African R4. Note the left hand side selector level above the pistol grip. R4 rifles have both the AK-style large selector lever on the right side in addition to this selector lever on the left. This double lever set up has also been used on some current Bulgarian designs and on the Serbian M21 5.56mm weapon.

FURTHER READING

AK47 – the Story of the People's Gun
Michael Hodges
Hadder & Stroughton Limited 2007

AK47 – the Weapon that Changed the Face of War
Larry Kahaner
John Wiley & Son Inc 2007

The Gun that Changed the World
Mikhail Kalashnikov/Elena Joly
Polity Press 2006

Greenhill Military Manuals: Kalashnikov
John Walter
Greenhill 2002

Kalashnikov Rifles and their Variations
Joe Poyer
North Cape Publications Inc 2004

Weapons of War Series – the AK47
Chris McNab
Spellmouth/Amberbooks 2001

Kalashnikov the Arms and the Man
Edward Ezell
Collector Grade Publications 2001

The AK47 Story: Evolution of the Kalashnikov Weapons
Edward Ezell
Stackpole Books 1986

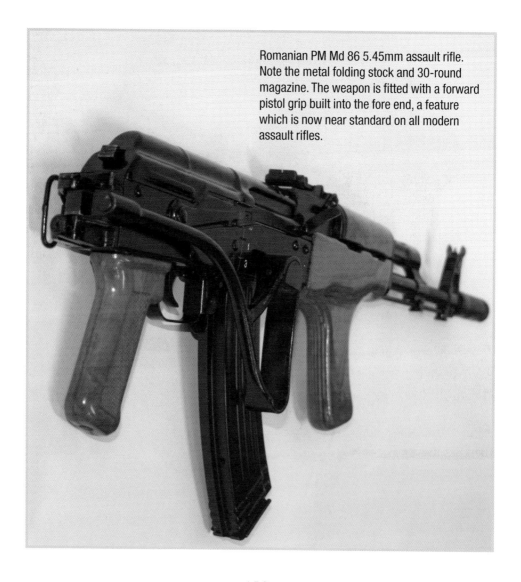

Romanian PM Md 86 5.45mm assault rifle. Note the metal folding stock and 30-round magazine. The weapon is fitted with a forward pistol grip built into the fore end, a feature which is now near standard on all modern assault rifles.

Bulgarian AK74

East German MPiAKS-74 with metal stock side folded to the righ hand side of the receiver. The sling is grey blue coloured and the plastic 30-round magazine is designed to take 5.45mm x 39mm ammunition.

Index

7N10, 147
7N22, 148
7N24, 148
Afghanistan, 7, 10, 19, 21–22, 42, 61, 81
AK
    assembly, 70–71
    cleaning, 31, 51, 64, 69–70, 127
    loading, 64–66, 68
    magazine, 68, 69, 70
    safety, 68
    sight setting, 69
    stoppages, 69
    unloading, 67–68
    zeroing, 71–72
AK47
    invention, 15, 16, 17
    origins, 13, 14
    Soviet trials, 16, 17
    Type 1, 17, 23, 24, 25, 27, 34, 75
    Type 2, 17, 23, 24, 25, 27, 30
    Type 3, 17, 18, 23, 24, 25, 28, 29, 34, 74, 77, 87, 89, 92, 101
AK-55, 101
AK-63D, 29, 87, 103, 104
AK-63E, 29, 87, 107, 106, 107
AK74, 37, 44
    compensator, 18, 41, 43
    sling, 45
AK74M, 124
AK103, 19, 42, 124, 146, 149
AKK, 77
AKM, 33, 37
    bolt, 58
    bolt carrier, 57
    gas tube, 53
    recoil mechanism, 57, 54
    weight, 34
AKM-63, 101
Albania, 74, 75
Algeria, 73, 74
America, *see* US
Ammunition, 143

M43 round, 9, 14, 15, 17, 19, 20, 146,147
M74 round, 18, 19, 37, 38, 42, 147
AR, 77
AR-F, 78
AR-1, 78
AR15, 15, 17
AR-1F, 78
AR-M1, 78
AR-M1F, 80, 149
AR-M7F, 80
Avtomat Kalashnikova Modernizirovanniya, *see AKM*
Berlin Wall,
BP round, *see 7N22*
BS round, *see 7N24*
Burma, 76
Bulgaria, 10, 18, 36, 73, 74, 75, 76, 77, 78, 79, 80, 146, 149
Cambodia, 81
China, 9–11, 18–21, 28–29, 31, 73, 75, 76, 81, 82, 83, 84, 85, 86, 108, 132
Cold War, 10, 21
Cuba, 87
DDR, *see German Democratic Republic*
DPRK, 19, 42, 87–91, 136
East Germany, *see German Democratic Republic*
Egypt, 90
Galil rifle, 11, 130
German Democratic Republic, 18, 42, 91, 92, 93, 94, 95, 96, 97, 98, 99, 100, 114, 118, 130, 132, 146
Hungary, 18, 101, 102, 103, 104, 105, 106, 107, 113

Iran, 10, 73, 108, 126
Iraq, 51, 108, 109, 110, 111, 138
Kalashnikov, Mikhail, 7, 10–11, 15–17, 19, 23, 33, 36, 37, 38, 41–42, 73, 124, 126, 132, 135, 149
KbK AK, 113
KbK AKM, 113
KbK AKMS, 113–114
KbK AKS, 113
KbKg, 113
KbK wz 1980, 114
KbK wz 1988, 114
KLF, 108
KLS, 108
KLT, 108
M16, 17, 19, 21, 28, 38, 71, 104, 143
M43 round, 9, 14–15, 17, 19–20, 146, 147
M70AB2, 125
M70B1, 109–110, 125–129, 141
M74 round, 18, 19, 37, 38, 42, 147, 148
M76 sniper rifle, 136
Magazines, 144–146
  AK47 Type 1, 24
  AK47 Tyle 3, 25
  AK-63E, 105
  AK74, 39
  AK74 Type 98, 90
  AKM, 34, 35, 53
  M76, 137, 138
  M70B1, 127, 129
  MPiAK-74, 93
  MPiAKS-74n, 101
  PM Md 86, 118, 123
  PSL, 139
  SAR21, 130

INDEX

STG 942, 132
M76, 137, 138
M76 sniper rifle, 127, 136–138
MAZ, 126
MG 42, 14
Misr AKM, 91
Misr AKMS, 91
Model 89, 74
Model 89-1, 74
MPiAKS-74, 97
MPiAKS-74n, 97, 99, 101 MPiK, 91–93
MPiKS, 92
MPiKM, 92, 95, 97, 101
MPiKMS-72, 93, 96, 98, 99, 100, 118
MPiAK-74, 93, 118
MPiAK-74n, 97
NATO, 7, 9, 10, 11, 19, 20, 73, 112, 149
North Korea, see DPRK
North Vietnamese Army, 21, 28 see also Vietnam War
Pakistan, 112
PM Md 63, 115
PM Md 65, 115, 117
PM Md 86, 117, 118, 119, 120, 121–123, 145
PM Md 90, 122, 123
Poland, 18, 42, 112, 113, 114
PP round, see 7N10
PPS 43, 14
PPSH 41, 14
PSL sniper rifle, 139–142
QBZ 95, 132
R4, 130
RK95, 130
Romania, 115
Russia, 10, 14, 18, 19, 20, 42, 73, 90, 124, 146, 149 see also Soviet Union
SAR21, 130, 131
Second World War, 8, 13, 20, 70, 135
Serbia, 125, 126, 128, 129, 136, 150
SKS carbine, 14, 16, 23
Soviet Union, 7, 10, 17, 18, 20, 21, 28, 42, 74, 81, 90, 149
STG44, 14, 16
STG 940, 130
STG 942, 130, 132–133
Sudan, 73, 126
SVD sniper rifle, 135, 139
Tabuk sniper rifle, 139, 141
Type 56, 18, 28, 29, 74, 76, 81, 83, 108, 126, 129
Type 56-1, 11, 74, 82, 83
Type 56-2, 84, 85, 86
Type 58, 87, 89
Type 68, 87, 90
Type 98, 89, 90
Type A, 74
Type B, 74
US, 7, 10, 15, 17, 18, 19, 20, 21, 22, 38, 109, 130, 143
USSR, see Russia and Soviet Union
Venezuela, 19
Warsaw Pact, 18, 42, 73, 91, 130
Viet Cong, 21, 28, 71
Vietnam, 7, 9, 19, 20, 21, 22, 61, 71, 129
Vietnam War, 11, 19, 20, 21, 28, 87, 143
Yugoslavia, 113, 125, 135, 136, 139